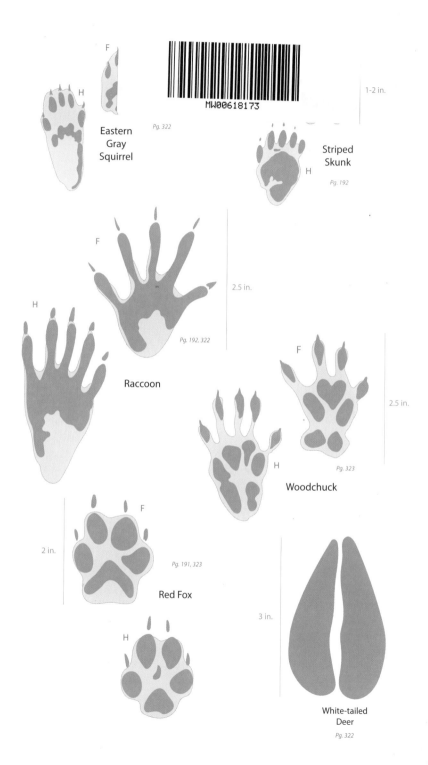

F

H

Eastern
Gray
Squirrel

Pg. 322

1-2 in.

Striped
Skunk

Pg. 192

H

F

H

2.5 in.

Raccoon

Pg. 192, 322

F

H

2.5 in.

Woodchuck

Pg. 323

F

2 in.

Red Fox

Pg. 191, 323

H

3 in.

White-tailed
Deer

Pg. 322

A Field Guide to

CAPE COD

Black Cherry tree on the dunes, Sandy Neck, Barnstable, Cape Cod.

A Field Guide to

CAPE COD

Including Nantucket, Martha's Vineyard, Block Island,
& Eastern Long Island

PATRICK J. LYNCH

All illustrations, maps, & photography by
the author unless otherwise noted

Yale
UNIVERSITY
PRESS

To Susan, Alex, Devorah, and Tyler,
to my good friend Frank Gallo,
and to the late Noble Proctor,
teacher, mentor, and friend

coastfieldguides.com

Yale University Press books may be purchased in quantity for
educational, business, or promotional use. For information, please
e-mail sales.press@yale.edu (US office) or sales@yaleup.co.uk (UK
office).

Designed by Patrick J. Lynch.

Printed in China.

ISBN 978-0-300-22615-7

Library of Congress Control Number: 2018937839

This paper meets the requirements of ANSI/NISO Z39.48-1992
(Permanence of Paper).

10 9 8 7 6 5 4 3 2 1

For the animal shall not be measured by man. In a world older and more complete than ours they move finished and complete, gifted with extensions of the senses we have lost or never attained, living by voices we shall never hear. They are not brethren, they are not underlings; they are other nations, caught with ourselves in the net of life and time, fellow prisoners of the splendor and travail of the earth.

—Henry Beston

Pitch Pines on the Bay View Trail, Wellfleet Bay Wildlife Sanctuary, Wellfleet.

CONTENTS

American Bullfrog at Trustom Pond National Wildlife Refuge on the coast of western Rhode Island, opposite Block Island.

During many trips to Cape Cod and the Outer Lands over the past 40 years I have benefited from the good company and deep birding and natural history expertise of Frank Gallo. Thanks, Frank, for all the great times and for always remembering to hit the Portuguese Bakery for goodies before our whale-watching trips.

I'd like to thank Patrick Comins, executive director of Connecticut Audubon, for his great advice and thorough review on my *Field Guide to Long Island Sound*, much of which also informs this book.

I thank Ralph Lewis, professor of geology at the University of Connecticut Avery Point campus, and former State Geologist of Connecticut. Ralph went above and beyond in sharing his expertise on New England's complex geologic history. Professor of geology J. Bret Bennington of Hofstra University generously gave me permission to use his excellent digital elevation map of Long Island. Geoimaging expert Paul Illsley graciously allowed me to use his magnificent bathymetry map of the Gulf of Maine.

I offer particular thanks to Jean Thomson Black, executive editor for life sciences at Yale University Press, for her faith in my work over the years and for being my constant advocate at the Press. I also thank the manuscript editor on this project, Laura Jones Dooley, for her wisdom, expertise, and guidance on almost every page here.

Last, and most of all, I thank my teacher, mentor, and friend, the late Noble Proctor, for his 43 years of wise counsel, for countless days of great birding and whale watching, and for introducing me and so many other Southern Connecticut State University students to the natural wonders of Cape Cod. I know that I and Noble's hundreds of friends throughout the world miss his good humor, sharp eyes, and awesome breadth of knowledge about the natural world. This book would not exist without Noble's wisdom and support.

PATRICK J. LYNCH

North Haven, Connecticut

coastfieldguides.com
@patrlynch
https://www.facebook.com/patrick.lynch1
patrlynch1@gmail.com

X

A young Herring Gull over a feeding Humpback Whale at Stellwagen Bank, off the northern tip of Cape Cod. The gulls feed on fish scraps left by the whales and are in little danger.

This book is a general introduction to the natural history of the ocean-facing coasts of southeastern New England and Long Island, with an emphasis on environments, not on particular locations. Although my focus is on the plants, animals, and physical foundations of this region, you cannot write about the natural world these days without constant reference to the effects of humanity and anthropogenic climate change. We live in the Anthropocene Epoch: human activity has become the dominant force that shapes our physical and biological environment.

The geologic and human history of our region also reminds us that we live on shifting ground. Sea level rise and shifting coastlines are nothing new, but the accelerating pace of climate change in the past 50 years has altered both our shorelines and the life around the region. Many of our southeastern New England lobster fisheries are dwindling because the waters are too warm for the Northern Lobster. Many formerly abundant food fish like the Atlantic Cod are endangered due to overfishing. Formerly southern birds like Turkey Vultures and Black Vultures are year-round residents, and the rising waters of the Atlantic not only shrink the habitats of beach-nesting birds like the Piping Plover but threaten the salt marsh meadows that are the breeding habitat for the Saltmarsh Sparrow and other endangered species.

This guide cannot be an exhaustive catalog of everything that lives in or near the shores of this region—such a book would be neither practical as a field guide nor very useful to the typical hiker, birder, kayaker, fisher, or boater. Here I have emphasized the most dominant and common plants and animals, plus a few interesting rarities like the Snowy Owl and locally threatened species like the Least Tern and the Piping Plover. My intent is to show you the major plants and animals that populate our shorelines and waters, so that you can walk into a salt marsh or onto a beach and be able to identify most of what you see, the first step in developing a deeper, more ecological understanding of the unique and beautiful aspects of the Outer Lands' major environments.

Useful companions to this guide

For readers interested in more information on the human history and environmental challenges facing this region, I highly recommend John T. Cumbler's *Cape Cod: An Environmental History of a Fragile Ecosystem* as a companion to this guide. No one has written better on the soul of Cape Cod than Henry Beston in *The Outermost House*, but Robert Finch's recent *The Outer Beach: A Thousand-Mile Walk on Cape Cod's Atlantic Shore* comes very close.

For recommendations on detailed field guides to specific topics such as plants, wildflowers, geology, birding, insects, and other wildlife, please consult the Bibliography.

Boston
Harbor

MASSACHUSETTS
BAY

Stellwagen Basin

STELLWAGEN
BANK

Cedar Pt.

Race Point

Provincetown

Duxbury Bay

Long Point

Truro

PLYMOUTH BAY

Plymouth

Wellfleet

Manomet
Point

CAPE COD
BAY

Billingsgate
Shoal

Eastham

Cape Cod Canal

CAPE COD

Nauset
Beach

Sandy Neck

Pleasant
Bay

Barnstable

Chatham

Chatham
Harbor

Chatham
Roads

BUZZARDS
BAY

Hyannis
Port

Monomoy
Island

Bearse
Shoals

Woods Hole

Horseshoe
Shoal

Pollock Rip
Channel

Nobska Pt.

Naushon Is.

NANTUCKET
SOUND

ELIZABETH ISLANDS

Vineyard
Haven

Oak Bluffs

Cape Pogue

Great Point

Great Round
Shoal Channel

Cuttyhunk Is.

VINEYARD
SOUND

Edgartown

Chappaquiddick Is.

Nantucket

Gay Head

Martha's
Vineyard

Muskeget Is.

Wasque
Shoal

Tuckernut Is.

Nantucket

Siasconset

Southwest
Shoals

Old Man
Shoal

Noman's Land Island
National Wildlife Refuge

NANTUCKET
SHOALS

Old South
Shoal

0   5   10   15   20   25   30
MILES

North

0   10   20   30   40   50
KILOMETERS

Mercator projection. Depth markers in feet.

CHESAPEAKE BAY

New York City

Fire Island

Long Island

LONG ISLAND
SOUND

Montauk
Point

Orient
Point

BLOCK ISLAND
SOUND

Block Island

NARRAGANSETT BAY

BUZZARDS
BAY

Cape Cod

Plymouth

CAPE COD BAY

Race Point

NASA Image

INTRODUCTION

Stellwagen Bank, one of the world's best whale-watching areas, is a short distance off the northern tip of Cape Cod. Here a Humpback Whale rolls its tail at the beginning of a deep dive.

This is a guide to the natural history of the Outer Lands, the sandy, ocean-facing islands and peninsulas that stretch in a 250-mile arc across southeastern New York and New England from Rockaway Beach on Long Island, encompassing Block Island, the sandy western Rhode Island coast, Martha's Vineyard, and Nantucket and its smaller out-islands, to Race Point at the north end of outer Cape Cod. It may seem surprising to unite the urbanized beaches of southwestern Long Island with remote places like Nantucket, the "far away land," as Native Americans called it. But the salty, windswept outlands that front the Atlantic Ocean have a remarkable amount in common, both in their origin and glacial geology and in the unique communities of plants and animals that make the Outer Lands feel distinctive and unified, whether you take the subway to Rockaway Beach or ride the Boston–Provincetown fast ferry to the Provincelands.

The Outer Lands
The unifying concept of the Outer Lands as a distinct geographic and ecological region is widely appreciated today by geographers and in the past by both the original Native American inhabitants and the later American sailors, whalers, fishermen, and tradesmen whose sense of their distinct regional culture owed little to either the farmers of interior Connecticut and Massachusetts or the Brahmins of Boston. The term "Outer Lands" was further popularized 50 years ago by author Dorothy Sterling, whose delightful natural history of *The Outer Lands* I highly recommend. A revised edition of Sterling's book is in print and easily obtained.

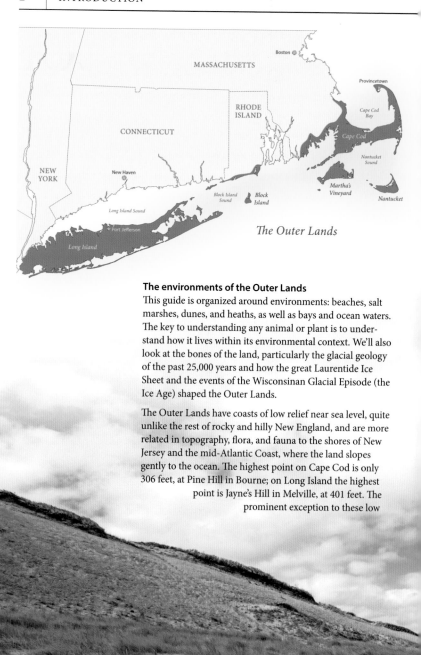

The Outer Lands

The environments of the Outer Lands

This guide is organized around environments: beaches, salt marshes, dunes, and heaths, as well as bays and ocean waters. The key to understanding any animal or plant is to understand how it lives within its environmental context. We'll also look at the bones of the land, particularly the glacial geology of the past 25,000 years and how the great Laurentide Ice Sheet and the events of the Wisconsinan Glacial Episode (the Ice Age) shaped the Outer Lands.

The Outer Lands have coasts of low relief near sea level, quite unlike the rest of rocky and hilly New England, and are more related in topography, flora, and fauna to the shores of New Jersey and the mid-Atlantic Coast, where the land slopes gently to the ocean. The highest point on Cape Cod is only 306 feet, at Pine Hill in Bourne; on Long Island the highest point is Jayne's Hill in Melville, at 401 feet. The prominent exception to these low

seascapes are the sand and clay cliffs of Montauk, Mohegan Bluffs on Block Island, the colorful cliffs of Aquinnah on Martha's Vineyard, and the 18-mile stretch of cliffs above the Outer Beach of Cape Cod. But as dramatic as the cliffs can be close up, none rise more than 150 feet above the sea, and all are composed of soft sands, silts, and clays whose edges lose on average three horizontal feet per year to the waves of the winter Atlantic seas.

The Outer Lands sit within the larger context of the Gulf of Maine, the New York Bight, and the Western Atlantic Ocean. The Hudson Canyon south of Long Island, the Nantucket Shoals, Georges Bank, and Stellwagen Bank are justly famous for the rich variety of sea life they harbor and sustain, even after centuries of overfishing and whaling. In deeper waters at the edge of the continental slope, a New England Coral Canyons and Seamounts Marine Sanctuary has been proposed

The earthen cliffs above Lecount Hollow Beach, Wellfleet. The cliffs here erode back about three feet per year, mostly due to winter storms.

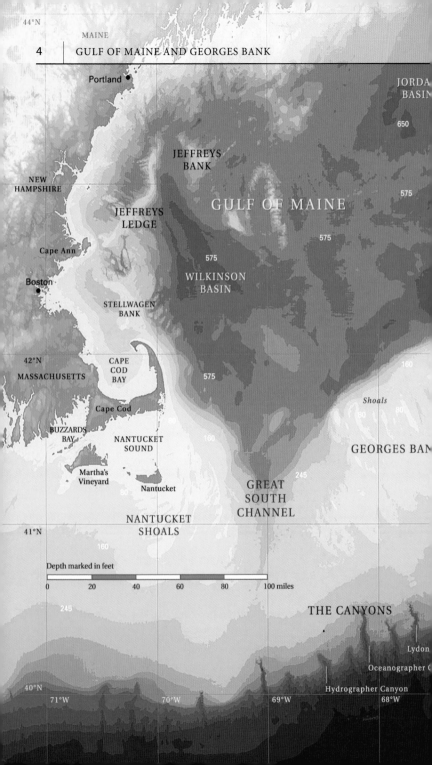

44°N

Portland

JORDAN
BASIN

650

JEFFREYS
BANK

NEW
HAMPSHIRE

GULF OF MAINE

575

JEFFREYS
LEDGE

575

Cape Ann

575

Boston

WILKINSON
BASIN

STELLWAGEN
BANK

160

42°N
MASSACHUSETTS

CAPE
COD
BAY

575

Cape Cod

80

BUZZARDS
BAY

NANTUCKET
SOUND

160

Shoals

80   80

GEORGES BANK

Martha's
Vineyard

Nantucket

80

245

GREAT
SOUTH
CHANNEL

NANTUCKET
SHOALS

41°N

160

Depth marked in feet

0   20   40   60   80   100 miles

245

THE CANYONS

Lydon

Oceanographer C

40°N

Hydrographer Canyon

71°W   70°W   69°W   68°W

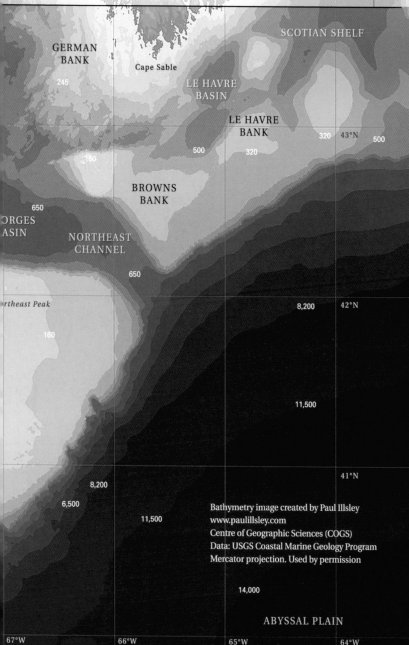

NOVA SCOTIA
44°N
575
Yarmouth

SCOTIAN SHELF

GERMAN
BANK
245
Cape Sable

LE HAVRE
BASIN

LE HAVRE
BANK
500   320   320   43°N   500

180

BROWNS
BANK

650

ORGES
ASIN

NORTHEAST
CHANNEL

650

8,200   42°N

rtheast Peak

160

11,500

41°N

8,200

6,500

11,500

Bathymetry image created by Paul Illsley
www.paulillsley.com
Centre of Geographic Sciences (COGS)
Data: USGS Coastal Marine Geology Program
Mercator projection. Used by permission

14,000

ABYSSAL PLAIN

67°W   66°W   65°W   64°W

for the famously productive and diverse canyons area around Oceanographer Canyon south of Georges Bank. Although many of these ocean features are remote from the shores of the Outer Lands, the abundant marine wildlife they sustain is directly connected to the region's sport fishing, whale watching, and commercial fishing, as well as what can be seen in the wrack line at the beach.

About terminology

Even the proudest Cape residents will admit that Cape Cod regional terminology is both confused and confusing. The local terms for parts of the Cape are a jumble of early colonial terms and old sailor slang, when "down" meant downwind and "lower" meant, well . . . whatever the speaker understood it to mean, regardless of how unhelpful the word might be to outsiders. So let's at least be consistent within these pages. I've laid out the regions of Cape Cod in the figure at right, mostly drawn from current usage in the Cape Cod real estate industry. When I refer to the Outer Lands I mean the whole region, as shown on page 2. When I'm referring just to Cape Cod and the nearby islands of Nantucket, Martha's Vineyard, and smaller islands like the Elizabeth Islands, I'll call them the Cape and Islands. This book covers only the environments of the southeastern part of New York's Long Island, including the barrier islands of the South Shore and Montauk Point, but for simplicity I'll just refer to Long Island.

Sunsets at Race Point Beach in the Provincelands rarely disappoint.

Warm summer winds
from the southwest

Downwind

Upwind

Outer
Cape

(also sometimes called
Lower Cape, but
not in this book)

Older terminology from the days of
sail was based on the predominant
southwesterly summer winds:
you went upwind, or "up Cape," and
downwind, or "down Cape"

"Going down Cape"

"Going up Cape"

Lower
Cape

Upper Cape

Mid-Cape

The dune Pitch Pine forest at Sandy Neck, Barnstable.

A Pitch Pine on the edge of the eroding cliffs above Marconi Beach, Wellfleet.

GEOLOGY OF THE OUTER LANDS

Of all the forces that come together to create a landscape, time is the hardest to grasp. Glacial ice, meltwater, ocean waves, and wind formed the surface geography of the Outer Lands, but to truly understand this landscape you must remember how young this land is, and how recently—at least by geologic standards—the dramatic events that shaped the Cape and Islands unfolded. If New England's 500-million-year geologic history were the equivalent of a 24-hour day, all the major events in the formation of the Cape and Islands landscape would take place in the last five seconds before midnight.

Twenty thousand years ago the land that became Cape Cod was a low ridge of glacial rubble several hundred miles north of the Atlantic Coast, supporting sparse patches of arctic tundra vegetation. Half-buried chunks of ice as big as city blocks were scattered across the terrain, melting over hundreds of years to become lakes and ponds. Today's Georges Bank was a huge forested peninsula that extended several hundred miles into the North Atlantic Ocean. Mountains of glacial ice 2,000 feet tall stood on the northern horizon. The sea level was more than 300 feet lower than today, and to the south of today's Outer Lands a vast expanse of northern forest and open tundra covered what is now the continental shelf. Mammoths, Mastodons, Dire Wolves, and giant Musk Oxen roamed a landscape that today lies deep under the ocean.

It takes imagination and a perspective far beyond the scale of a human lifetime to see the events of 20,000 years ago as both long ago and relatively sudden and recent. But knowing the geologic history of the Outer Lands may give you a useful

Dune ridge and an interdune swale at Sandy Neck, Barnstable. Large sand dunes like these are created primarily by strong winter winds that drive sand inland from beaches and sandspits. Dune sand is lighter in color and weight and finer in texture than beach sand because it is composed mostly of small grains of light-colored quartz that are easily moved by the wind.

Before the Quaternary glacial periods, New England had a coastal plain and low, sandy shores that resembled what we see along the mid-Atlantic Coast today. This view is from Currituck Sound, on the Outer Banks of North Carolina.

viewpoint on today's concerns about rising sea levels, a warming planet, and our rapidly changing coastal environment.

Forming the Cape and Islands

A defining characteristic of the Outer Lands is the lack of exposed bedrock or rocky shores. We know few details about the rocky underpinnings of the Outer Lands because the bedrock is buried so deeply beneath ancient coastal plain sediments, as well as by much more recent surface sediment layers created by multiple glacial periods. The bedrock in the Long Island and Cape Cod areas generally slopes toward the southeast. On Cape Cod the bedrock is on average about 300 feet below surface level, and on Nantucket the bedrock is between 1,500 and 1,800 feet below ground level. On the North Shore of Long Island the bedrock is about 300 feet deep, and under Fire Island the bedrock is over 1,500 feet below ground level.

The bedrock foundation

The Outer Lands' geologic history begins about 500–300 million years ago, when the process of plate tectonics brought together most of the world's ancient landmasses into a supercontinent called Pangaea. As the continents crushed together to form Pangaea, the bedrock that was much later to underlie New England and Long Island was heated, folded, and faulted into a complex series of north-south-oriented valleys and hills. This north-south pattern of hills and valleys would later play an important part in the development of rivers in New England and in the human history of the region. The enormous heat and stress of the continental collisions created or modified much of the exposed bedrock we see today along the northern coastline of Long Island Sound in Connecticut, the Rhode Island shoreline around Narragansett

Bay, and much of the coastline of the Gulf of Maine north of Plymouth, Massachusetts.

Pangaea existed for about 50 million years as a supercontinent and then began breaking up in a process that created North America, Africa, and the Atlantic Ocean. As tectonic forces pulled the North American plate away from what became Africa, a narrow ocean strait formed: the early Atlantic Ocean. The rifting (pulling apart) of Pangaea caused great tension stresses in the bedrock, and giant cracks (rift basins) formed along the eastern edge of the North American plate. Today's Hartford Basin, the great central valley of Connecticut and western Massachusetts, is one of those huge tension cracks in the bedrock of the eastern edge of the North American plate.

As the continental pieces of Pangaea broke up about 250 million years ago, the Appalachian Mountains became the eastern coastline of the newly formed North American continent, along the shores of the developing Atlantic Ocean. The Appalachians were tall and rugged, but over the next 200 million years weather and water eroded the peaks, and much of their former substance washed down to create the broad coastal plains of today's Atlantic coastline south of New York Harbor. Off the Atlantic Coast, layers of eroded sand and silt from the Appalachians also form much of the continental shelf. Most of New England once had a broad, gradually sloping coastal plain and sandy barrier islands similar to what we see today on the coast south of New York, but multiple glacial periods and rising seas over the past 2.6 million years have largely eroded, buried, or submerged the visible traces of New England's ancient coastal plain.

Until about 2.6 million years ago the major forces shaping the Atlantic coastal plain were the same weathering and stream erosion that wore down the Appalachians. During the Tertiary Period (66–2.6 million years ago) the sea level was often much lower than it is today, and river and stream erosion created the ancestral valleys of today's great coastal gulfs, bays, and inlets: the Gulf of Maine, Cape Cod Bay, Long Island Sound, Block Island Sound, the Hudson River Canyon southeast of New York Harbor, and the Chesapeake and Delaware Bays. The Great South Channel east of Cape Cod and the Northeast Channel at the east end of Georges Bank also originated as ancient river valleys during the Tertiary. (See illustration, pp. 4–5.)

Today's low, rounded Appalachian Mountains are the heavily eroded remnants of taller, more rugged peaks that once probably looked like these mountains in the modern Sierra Nevada. Much of today's Atlantic coastal plain and the continental shelf off the Outer Lands originated as sediments eroded from those ancient Appalachians. Layers of these ancient sediments also lie under Long Island, Block Island, Martha's Vineyard, Nantucket, and parts of Cape Cod.

The glacial periods

The past 2.6 million years have been marked by a series of glacial periods collectively known as the Quaternary or Pleistocene glaciations. The Cape Cod region and Long Island

Glacial ice is nothing like the clear, clean ice cubes in your freezer. Glaciers are full of rocks, giant boulders, sand, and fine silt. Here two modern glaciers (top and bottom of picture) on Washington State's Mount Rainier are so full of rock debris that you can hardly tell where the rock ends and the ice begins.

*When people refer to the Ice Age in New England, they usually mean the Wisconsinan Glacial Episode, 85,000–16,500 years ago, but the term "Ice Age" is ambiguous. There have been multiple ice ages in our region over the past 2.6 million years.

show evidence of at least four distinct Pleistocene glaciations. Glacial periods occur when a complex set of astronomical conditions (variations in the earth's axial tilt and orbital distance from the sun) combine with other general climatic and geologic factors to cause long-term climate cooling. As the earth grew colder, winter snows did not entirely melt away in summer, and as the snow accumulated over thousands of years the ice cap of the northern hemisphere expanded southward. Today we are in a relatively warm interglacial period, and our Arctic glaciers are remnants of the last glacial period.

The most recent glaciation in the Outer Lands region, the Wisconsinan Glacial Episode, began about 85,000 years ago and ended in southern New England about 16,500 years ago.* It is called the Wisconsinan because the first major studies of this glaciation were conducted in Wisconsin. The continent-sized glacier of the Wisconsinan Episode is called the Laurentide Ice Sheet, named for the Laurentide region of northeastern Canada, where the ice was thought to have originated.

At the peak of the Wisconsinan Episode 25,000 years ago, the Laurentide Ice Sheet blanketed New England (see illustration, pp. 16–17) and reached as far south as the middle of present-day Long Island. In places the ice sheet was thousands of feet thick, and the landscape of the Outer Lands region resembled central Greenland today (see p. 17). At its peak so much of the earth's water was bound up in glacial ice that the sea level was 400 feet lower than it is today, and a large area of dry land extended south of the present-day coasts of Long Island and New England. This ice-free land resembled the spruce taiga forests and tundra of northern Canada today and provided a refuge area—a refugium—where many of the plant and animal species in our area today were able to survive. These species began to repopulate the Outer Lands when the ice sheet started to retreat about 24,000 years ago.

The great glacial moraines and ice lobes

The position of a glacier's ice front is determined by the balance between ice supply and melting rates. At a higher rate of supply, the ice front advances across the landscape. At a higher melting rate, it retreats. Where ice supply and melting rates are in balance, the ice front remains stationary, known as a stillstand.

When the ice front maintains a stillstand, the glacier acts like a conveyor belt, dumping sand and rock debris along the melting ice front and creating a pile of stony, sandy rubble known as an end moraine. The Outer Lands region has two types of end moraines. The end moraine that marks the farthest advance of the ice lobes is the terminal moraine, formed

The Laurentide Ice Sheet. At its maximum extent during the Wisconsinan Glacial Episode 25,000 years ago, the Laurentide Ice Sheet, a single, massive glacier, covered most of northeastern, eastern, and north-central North America. It may be easiest to think of the Wisconsinan glacier as a giant extension of the polar ice cap.

about 25,000 years ago (see green line on illustration, pp. 20–21). In our region the terminal moraine of the Wisconsinan Episode ice sheet, the Ronkonkoma Moraine, formed the backbone of Long Island, as well as Montauk Point, Block Island, Martha's Vineyard, and Nantucket.

End moraines that mark melt-back positions north of the terminal moraine are called recessional moraines. A series of recessional moraines formed about 21,300 years ago in a temporary period of cooling. The ice lobes stopped melting northward and began readvancing southward. As they moved across the landscape they not only dumped new glacial drift* sediments but, in the Cape Cod area, also bulldozed up layers of older glacial sediments, creating a new line of moraines (see brown line on illustration, pp. 20–21). Most of the highlands of Upper Cape Cod, the Elizabeth Islands, and the North Shore of Long Island were created during this second ice lobe advance.

Beginning about 20,000 years ago the earth's climate began to warm very quickly. The Laurentide Ice Sheet melted back, and the Wisconsinan Glacial Episode effectively ended in the Outer Lands about 16,500 years ago.

*Glacial drift (sometimes also called glacial till) is the most common and widespread form of glacial deposit in the Outer Lands area. Glacial drift consists of a rough mixture of boulders, smaller stones, sand, silt, and clay particles. Drift is essentially a jumble of everything the ice sheet picked up as it scraped over the New England landscape: old sediment layers from New England's former coastal plain, as well as rocks derived from the bedrock of New England.

The Laurentide Ice Sheet over New England 25,000 years ago
When you consider how radically different the New England landscape was at the height of the Ice Age, 25,000 years doesn't seem so long ago. This region has seen enormous changes in geography, sea level, and climate in a relatively short time.

To clarify glacial terminology, the **Laurentide Ice Sheet** is the name of the vast glacier that once covered much of North America and was named for the Laurentian Mountain region of Canada. This illustration shows the maximum southern extent of the Laurentide Ice Sheet, about 25,000 years ago.

Geologists call this most recent glacial time period the **Wisconsinan Glacial Episode**. It started about 85,000 years ago, peaked 25,000 years ago, and ended 16,500 years ago.

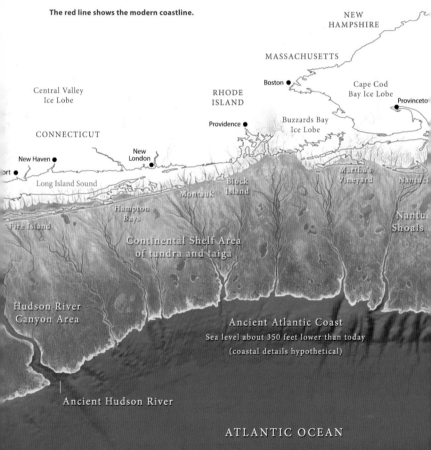

The red line shows the modern coastline.

NEW HAMPSHIRE

MASSACHUSETTS

Boston

Cape Cod
Bay Ice Lobe

Provinceto

Central Valley
Ice Lobe

RHODE
ISLAND

CONNECTICUT

Buzzards Bay
Ice Lobe

Providence

New
London

New Haven

ort

Long Island Sound

Block
Island

Martha's
Vineyard

Nantuc

Montauk

Hampton
Bays

Nantu
Shoals

Fire Island

Continental Shelf Area
of tundra and taiga

Hudson River
Canyon Area

Ancient Atlantic Coast
Sea level about 350 feet lower than today
(coastal details hypothetical)

Ancient Hudson River

ATLANTIC OCEAN

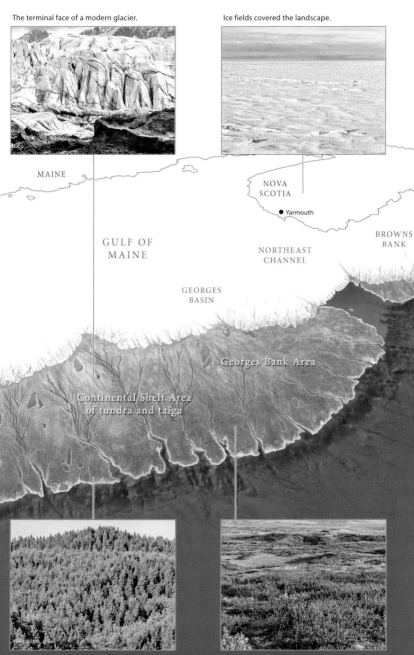

The terminal face of a modern glacier.

Ice fields covered the landscape.

MAINE

NOVA
SCOTIA

● Yarmouth

BROWNS
BANK

GULF OF
MAINE

NORTHEAST
CHANNEL

GEORGES
BASIN

Georges Bank Area

Continental Shelf Area
of tundra and taiga

Taiga forests of evergreens in the refugium
area south of the glaciers.

The refugium area was largely tundra, much
like the far north of Canada and Alaska
today.

About the dates in this chapter:
You may notice that the dates given here for geologic events in the past 25,000 years are different from those you might have seen in older texts. The dates here are derived from recent geologic studies that use surface exposure dating, in which the decay of radioactive compounds such as beryllium-10 is used to derive dates for glacial events.

The major ice lobes

The Laurentide Ice Sheet covering New England was divided into a number of distinct glacial flows called ice lobes. The shape and position of these lobes shaped the present geography of the Outer Lands.

Three ice lobes crossed the area that later became Cape Cod, Martha's Vineyard, Nantucket, and the Elizabeth Islands (see map, pp. 20–21). At the southernmost terminus of the ice sheet, the edges of the three ice lobes stood at the latitude of present-day Martha's Vineyard and Nantucket. Both islands were formed as roughly triangular piles of glacial debris and moraines in the angles between the ice lobes (see inset, p. 20). The Buzzards Bay Lobe formed much of the moraine highlands of western Martha's Vineyard at this time. During the cooling and readvance of the ice lobes about 21,300 years ago, the Buzzards Bay Lobe shoved up piles of older glacial debris layers that later became the Elizabeth Islands. The southernmost extension of the Cape Cod Bay Lobe created a terminal moraine that later became the eastern parts of Martha's Vineyard and the western portions of Nantucket (see green upland areas of islands, p. 21).

The colorful cliffs of Aquinnah
in western Martha's Vineyard are a mix of recent sediments from the Wisconsinan Glacial Episode 25,000 years ago and far older Cretaceous Era sediments—some as old as 60 million years—that were once the coastal ocean bottom along the ancient North American continent.

East of modern Cape Cod the huge South Channel Lobe filled the long, submerged ocean-bottom valley now known as the Great South Channel. As this lobe melted back, west-flowing glacial outwash streams ran off its heights and created most of the Outer Cape area, such as the outwash plains and river delta sediments of Truro, Wellfleet, and Eastham (see illustration, pp. 26–27).

Long Island was formed by several broad ice lobes that moved south across Connecticut and southern New York and stopped at about the east-west midpoint of today's Long Island. Montauk Point and the backbone highlands of Long Island mark the terminal moraine of the ice sheet. The northern shoreline of Long Island, Orient Point, Plum Island, and Fishers Island were formed about 21,300 years ago as a recessional moraine.

The formation of Martha's Vineyard, Nantucket, and Block Island

When the regional ice lobes reached their maximum southern stillstand positions, they created a roughly east-west line of glacial sediments and boulders that stretches from the New York City area to the eastern shore of Nantucket (see green line on map, pp. 20–21). South of what is now Cape Cod, the intersection of the Buzzards Bay and Cape Cod Bay ice lobes created a roughly triangular moraine in the notch between the lobes that later became the northern highlands and southern outwash plains of Martha's Vineyard. Similarly, in a triangular notch between the arcs of the Cape Cod Bay and South Channel lobes, a smaller raised moraine area grew that became the uplands of Nantucket. As the ice sheet melted back to the north, many streams of meltwater redistributed the raised moraine sediments southward, creating the outwash plains that make up most of both Nantucket and Martha's Vineyard. The large area of shallow shoals south of Nantucket originated as an outwash plain that was later submerged by rising seas.

On both Martha's Vineyard and Nantucket, the Wisconsinan glacial sediments sit atop older glacial deposits, layers of sediment from the ancient Atlantic coastal plain, and even older marine deposits from the continental shelf. The ice lobes of the Wisconsinan glacier dumped glacial sediments in the island areas and bulldozed far older sediments into a complex jumble of layers. In the colorful cliffs of Gay Head on Martha's Vineyard and in the Sankaty cliffs of eastern Nantucket, ancient Cretaceous Period (66 million years ago) sediments and later ocean-bottom sediments are intertwined with much younger glacial sediments.

The core of Block Island was created 25,000 years ago by the same terminal moraine system as Long Island. The Wisconsinan ice deposited glacial drift over much older sediments layers from earlier glacial periods, creating what later became the core moraine uplands of Block Island. Meltwater from the glaciers and thousands of years of normal weathering distributed the moraine sediments into small outwash plains. Marine erosion over the past 6,000 years has carved the famous Mohegan Bluffs cliffs at the southern end of Block Island.

The core moraines and outwash plains of Block Island, Martha's Vineyard, and Nantucket were all later heavily eroded and reshaped by ocean waves. All three islands are now substantially smaller than they were when the rising seas reached their approximate locations about 6,000 years ago.

The formation of Long Island

Long Island is often described as entirely the work of the Wisconsinan Glacial Episode, but much of the underlying structure of Long Island long predates the Wisconsinan ice, and in many places on the island the layers of Wisconsinan glacial sediments are a thin veneer over far older Atlantic coastal plain sediments.

The bedrock foundation of Long Island is similar to the schist and granite gneiss rock formations that underlie the southwestern coast of Connecticut and Westchester County, New York. From the northern coast of Long Island Sound the bedrock formation slopes downward toward the south: it is

PATRICK J. LYNCH

A Field Guide To

LONG ISLAND SOUND

For more details on the geology of Long Island, see *A Field Guide to Long Island Sound*, also from Yale University Press.

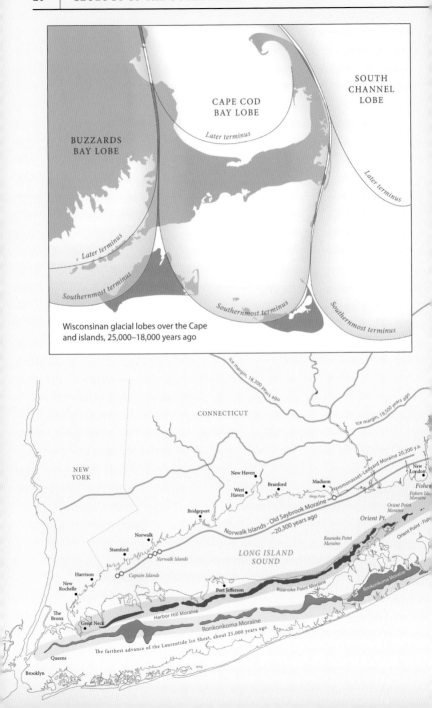

BUZZARDS
BAY LOBE

CAPE COD
BAY LOBE

SOUTH
CHANNEL
LOBE

Later terminus

Later terminus

Later terminus

Southernmost terminus

Southernmost terminus

Southernmost terminus

Wisconsinan glacial lobes over the Cape
and islands, 25,000–18,000 years ago

Ice margin, 18,300 years ago

Ice margin, 19,500 years ago

CONNECTICUT

NEW
YORK

New Haven
West
Haven
Branford
Madison
New
London
Hammonasset–Ledyard Moraine 20,200 y.a.
Fishe

Mega Point

Fishers Isle
Moraine

Bridgeport

Norwalk Islands – Old Saybrook Moraine
~20,300 years ago

Orient Pt.
Orient Point - Fis

Orient Point
Moraine

Norwalk

Stamford

Norwalk Islands

LONG ISLAND
SOUND

Roanoke Point
Moraine

Harrison

Captain Islands

New
Rochelle

Ronkonkoma Moraine

The
Bronx

Port Jefferson

Roanoke Point Moraine

Great Neck

Harbor Hill Moraine

Ronkonkoma Moraine

Queens

The farthest advance of the Laurentide Ice Sheet, about 25,000 years ago

Brooklyn

Montauk Point on Long Island is part of the same moraine system that created Martha's Vineyard and Nantucket.

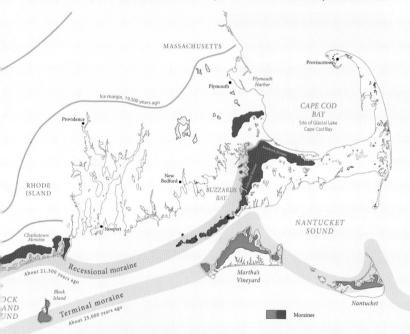

The great regional moraines of southern New England and Long Island

The green line marks the southernmost edge of the Laurentide Ice Sheet that covered New England 25,000 years ago. The terminal moraine is a huge pile of boulders, rocks, sand, and silt that today forms much of the outer New England and Long Island coastline.

The Laurentide Ice Sheet did not melt at a steady rate. About 21,300 years ago, the climate cooled again, and for thousands of years the ice sheet piled up a second massive recessional moraine system, the brown line on the map.

100 feet or more below the North Shore of Long Island and 2,000 feet deep under Fire Island and the southern shores of Long Island.

The core structure of Long Island consists primarily of sediment layers that date from the Cretaceous Period, 145–66 million years ago. These are sediments eroded from the Appalachian Mountains, the same sediments that today make up much of the coastal plain and continental shelf of the Atlantic coastline south of New York Harbor. During the Tertiary Period, 66–2.6 million years ago, multiple changes in sea level modified these Cretaceous strata in ways that are not thoroughly understood, but today the strata are well known to the general public as the primary source of Long Island's drinking water. The Magothy, Raritan, and Lloyd aquifers are water-rich layers of Cretaceous Period sediments that form Long Island's foundation.

The glacial outwash plains of Long Island began to form when the major regional moraines were formed. Meltwater streams and thousands of years of weathering distributed the lighter sand, silt, and gravel of the moraines into broad, flat outwash plains to the north and south of the Ronkonkoma Moraine (see map, below). The Harbor Hill–Roanoke Point recessional moraine also weathered into outwash plains south of the moraine, but to the north the waves of Long Island Sound have steadily cut into the moraine face, creating the almost continuous line of earthen cliffs that form the North Shore. About 6,000 years ago the rising sea level reached the

An elevation map of Long Island, showing the two major moraines that formed the island, the Ronkonkoma and Harbor Hill–Roanoke Point Moraines. Note the large, smooth outwash plains that lie north and south of the Ronkonkoma Moraine. The Ronkonkoma is a terminal moraine, formed at the southernmost edge of the glacier 25,000 years ago. The smaller Harbor Hill–Roanoke Point–Orient Point moraine complex was formed about 18,000 years ago, when the ice sheet had melted back 8–10 miles to the north and formed a second recessional moraine.

Digital elevation map created by Prof. J. Bret Bennington of Hofstra University and used with permission.

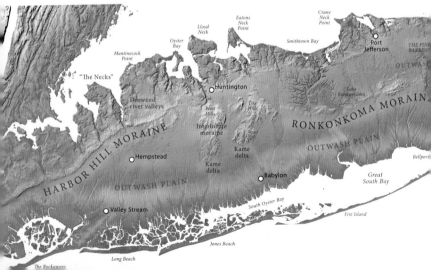

area that is now the south coast of Long Island, and the ocean waves began to erode the soft glacial drift of the outwash plains and to build the barrier islands and wide, sandy ocean beaches that have made southeastern Long Island and the Montauk Point area a famous summer vacation destination.

The formation of Cape Cod

Cape Cod, the youngest glacial feature of the Outer Lands, was shaped thousands of years after Martha's Vineyard, Nantucket, and most of Long Island. Cape Cod was formed by three major events: the movements and later melting of the Wisconsinan glacier, rising sea levels in postglacial times, and erosion and deposition of the glacial sediments by ocean waves.

From about 24,000–22,000 years ago the earth's climate was rapidly warming, and the great ice sheets of the Wisconsinan glaciation had melted back from their maximum southern positions. About 21,300 years ago the Cape Cod Bay Lobe of the Laurentide Ice Sheet had melted back to a position about where the southern shore of Cape Cod Bay is today (see

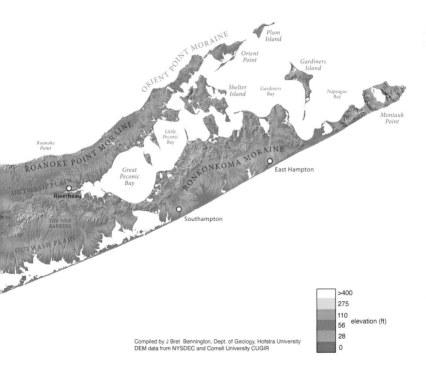

Compiled by J Bret Bennington, Dept. of Geology, Hofstra University
DEM data from NYSDEC and Cornell University CUGIR

brown line on map, pp. 20–21). Then for some hundreds of years the climate cooled, so much so that the Cape Cod Bay and Buzzards Bay Lobes began to advance southward again. As the ice lobes moved south once more, they dug into the layers of glacial sediment in what is now the Upper Cape area, shoving the layers of glacial drift ahead of them like a giant bulldozer, creating the highlands we now know as the Sandwich and Buzzards Bay Moraines.

The climate then warmed again, the ice lobes melted back, and as the ice melted, a huge freshwater lake—Glacial Lake Cape Cod Bay—formed in the basin now occupied by modern Cape Cod Bay (see illustration, pp. 26–27). In the early stages of lake formation the water level of Glacial Lake Cape Cod Bay was 50–60 feet higher than the present sea level, and layers of mud and sand sediments from the lake bottom covered earlier glacial drift layers, creating the deep, nutrient-rich, and relatively stone-free soils that stretch along the bay coastline of the Cape from Sandwich to Eastham.

The coastal flatlands created by the lake sediments were ideal for European-style farming and animal husbandry, and the shores of Cape Cod Bay were some of the first Cape areas settled by Pilgrim and Puritan communities. The town centers of Sandwich and Barnstable sit on layers of lake-bottom sediments. Glacial Lake Cape Cod Bay existed for hundreds of years but gradually drained away to the ocean through outlet channels in the valley now occupied by the Cape Cod Canal and perhaps other smaller outlet streams.

Most of the Upper Cape, Mid-Cape, and Lower Cape areas consist of low and relatively smooth outwash plains of sand, silt, and rock material eroded from the Sandwich and Buzzards Bay Moraines. When the ice lobes retreated northward, iceberg-sized chunks of glacial ice were left behind, partially buried in glacial drift. As these huge chunks of ice melted, the surface of the Cape's outwash plains became pitted with small lakes, the famous kettle ponds of Cape Cod.

The origins of the Outer Cape area

The outer arm of the Cape is also an outwash plain, formed around 21,000 years ago by a complex interaction of the South Channel Lobe, the Cape Cod Bay Lobe, and Glacial Lake Cape Cod Bay (see illustration, pp. 26–27). The South Channel Lobe occupied most of the large undersea valley east of modern Cape Cod called the Great South Channel (see map, p. 4). The South Channel Lobe was far larger than the other ice lobes that formed Cape Cod and persisted much longer before it melted out of the Cape Cod region. The glacial outwash sediments that form the Outer Cape between

Dennis's Scargo Tower is a 30-foot cobblestone turret that sits atop one of the higher hills near the Sandwich Moraine. From the tower you get a nice overview of the rolling Mid-Cape moraine landscape and the classic kettle pond just below the tower.

The Sandwich and Buzzards Bay ice-thrust moraines

About 21,300 years ago the climate cooled, the ice sheet advanced south again like a giant bulldozer, and the forward edge of the glacier crumpled older layers of sediments into the raised Sandwich and Buzzards Bay Moraines. The moraines of Martha's Vineyard are also at least partially ice-thrust moraines.

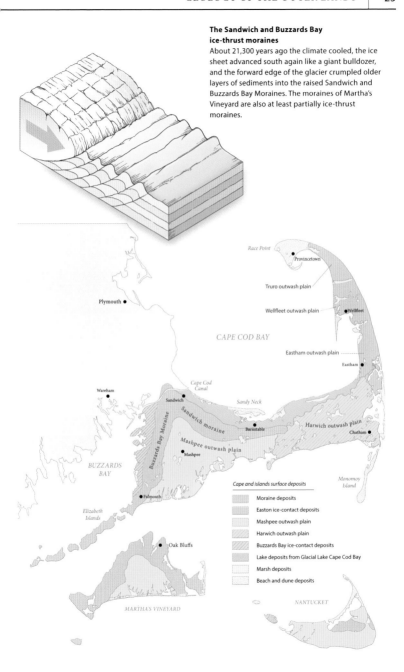

Race Point

Provincetown

Truro outwash plain

Plymouth

Wellfleet outwash plain

Wellfleet

CAPE COD BAY

Eastham outwash plain

Eastham

Cape Cod Canal

Wareham

Sandwich

Sandy Neck

Sandwich moraine

Barnstable

Harwich outwash plain

Chatham

Buzzards Bay Moraine

Mashpee outwash plain

Mashpee

BUZZARDS BAY

Monomoy Island

Falmouth

Elizabeth Islands

Oak Bluffs

Cape and islands surface deposits

Moraine deposits

Easton ice-contact deposits

Mashpee outwash plain

Harwich outwash plain

Buzzards Bay ice-contact deposits

Lake deposits from Glacial Lake Cape Cod Bay

Marsh deposits

Beach and dune deposits

NANTUCKET

MARTHA'S VINEYARD

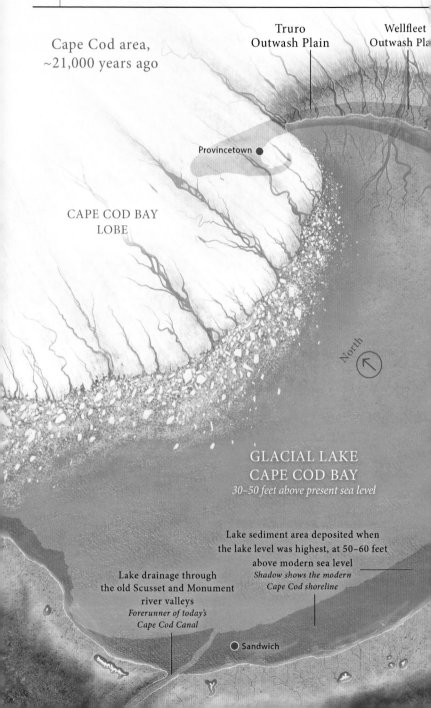

Cape Cod area,
~21,000 years ago

Truro
Outwash Plain

Wellfleet
Outwash Pla

Provincetown

CAPE COD BAY
LOBE

North

GLACIAL LAKE
CAPE COD BAY
30–50 feet above present sea level

Lake sediment area deposited when
the lake level was highest, at 50–60 feet
above modern sea level
*Shadow shows the modern
Cape Cod shoreline*

Lake drainage through
the old Scusset and Monument
river valleys
*Forerunner of today's
Cape Cod Canal*

Sandwich

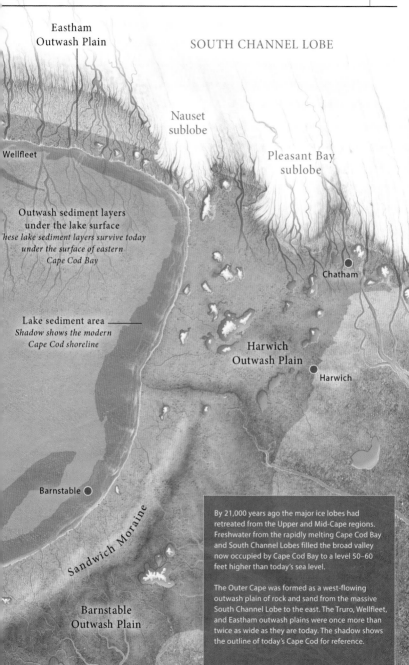

Eastham
Outwash Plain

SOUTH CHANNEL LOBE

Nauset
sublobe

Pleasant Bay
sublobe

Wellfleet

Outwash sediment layers
under the lake surface
*These lake sediment layers survive today
under the surface of eastern
Cape Cod Bay*

Chatham

Lake sediment area
*Shadow shows the modern
Cape Cod shoreline*

Harwich
Outwash Plain

Harwich

Barnstable

Sandwich Moraine

Barnstable
Outwash Plain

By 21,000 years ago the major ice lobes had
retreated from the Upper and Mid-Cape regions.
Freshwater from the rapidly melting Cape Cod Bay
and South Channel Lobes filled the broad valley
now occupied by Cape Cod Bay to a level 50–60
feet higher than today's sea level.

The Outer Cape was formed as a west-flowing
outwash plain of rock and sand from the massive
South Channel Lobe to the east. The Truro, Wellfleet,
and Eastham outwash plains were once more than
twice as wide as they are today. The shadow shows
the outline of today's Cape Cod for reference.

High Head in Truro south to Orleans came almost entirely from the South Channel Lobe.

The outwash plains of Outer Cape Cod consist primarily of sand, silt, and gravel sediments that washed down in a westerly direction from the heights of the South Channel Lobe east of today's Outer Cape. This westerly flow from the heights of the South Channel Lobe down toward Cape Cod Bay is still reflected in today's Outer Cape landscape, which tilts downward to the west from the high eastern ocean cliffs to the much lower Cape Cod Bay shoreline.

The Truro, Wellfleet, and Eastham outwash plains formed in the angle between the South Channel Lobe and the Cape Cod Bay Lobe (see illustration, pp. 26–27). Much of the outwash sediment flowed into Glacial Lake Cape Cod Bay. These lake sediments persist in modern Cape Cod Bay and explain why the Bay is so much shallower on its eastern side.

The hollows of the Outer Cape

Crossing the outwash plains of the Outer Cape are a series of steep-walled valleys, called hollows, that are much too large to have been cut by the tiny, slow streams that occupy some of them today. These valleys generally run east-west, and they slope down toward the Cape Cod Bay side, as do the higher plains around them. Geologists call these valleys outwash channels, and they were likely formed by meltwater streams that ran off the huge South Channel Lobe east of today's oceanside Outer Cape beaches (see illustration, pp. 26–27). Lecount Hollow, Cahoon Hollow, and Newcomb Hollow are all examples of outwash channels.

The view from Truro's Bearberry Hill, facing southeast toward the hollow of Ballston Beach in the cliffs of the Outer Cape. Note the broad, sandy overwash fan. In recent years winter storms have regularly overtopped Ballston Beach, creating the large overwash fan of sand stretching west from the beach. In the coming decades rising seas may spill over the Ballston Beach and Pamet River areas, perhaps eventually creating a small saltwater channel in the valley, effectively making the north tip of the Cape an island.

Ballston Beach

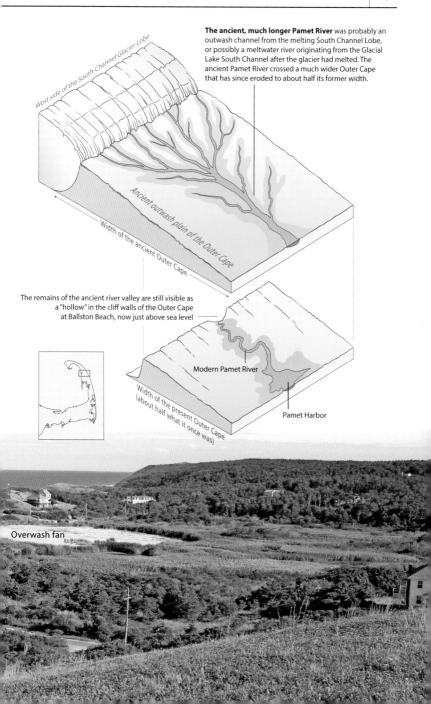

The ancient, much longer Pamet River was probably an outwash channel from the melting South Channel Lobe, or possibly a meltwater river originating from the Glacial Lake South Channel after the glacier had melted. The ancient Pamet River crossed a much wider Outer Cape that has since eroded to about half its former width.

West side of the South Channel Glacier Lobe

Ancient outwash plain of the Outer Cape

Width of the ancient Outer Cape

The remains of the ancient river valley are still visible as a "hollow" in the cliff walls of the Outer Cape at Ballston Beach, now just above sea level

Modern Pamet River

Width of the present Outer Cape (about half what it once was)

Pamet Harbor

Overwash fan

The Ballston Beach access area sits in a low notch between high sea cliffs. Six thousand years ago Cape Cod extended at least two miles to the east of the current beach, and the Pamet River was once at least twice as long as it is today. Erosion of the ocean face of the Cape has removed miles of land from the ocean side of the Outer Cape and has left the ancient river valley hollow as a notch in the cliffs, with the tiny, sluggish Pamet River running west toward Cape Cod Bay. At Ballston Beach the beginning of the Pamet almost meets the beach because most of the river's original headwater sources have been eroded away (see illustration, p. 29).

The Pamet River valley is the largest of the Outer Cape's hollows, and geologists debate the exact processes that formed it. The valley is probably a conventional outwash channel, formed by floods of westward-flowing meltwater from the now-vanished glacial South Channel Lobe. Or perhaps the valley was cut at least in part by centuries of underground meltwater springs flowing west from the South Channel Lobe in a process geologists call spring-sapping (discussed later in this chapter).

Sea level rise

At the height of the Wisconsinan glaciation 25,000 years ago the sea level was more than 400 feet lower than it is today, mostly because so much of the earth's water was bound up in ice sheets. It's important to remember this when discussing the formation of the Outer Lands, because all of Cape Cod, Martha's Vineyard, Nantucket, and Long Island began as inland locations far north of the Atlantic coastline. Not until about 6,000 years ago did ocean waves begin eroding and reshaping the Outer Lands shoreline areas we recognize today (see illustrations, p. 31).

The west-flowing Pamet River in Truro is barely a true river these days, and consists mostly of a small, spring-driven water channel that grades into a long, brackish and saltwater marsh as it nears the coast of Cape Cod Bay at Truro Harbor.

After the glacial maximum, the sea level rose gradually as the climate warmed and the Laurentide Ice Sheet began to melt. About 15,000 years ago the earth's climate began to warm at a much faster rate, and a series of meltwater pulses caused a rapid rise in sea levels until about 8,000 years ago, when the rate of sea level rise slowed significantly. Thus the sea level has been steadily rising for the past 25,000 years. Today's concern about rising sea levels centers mainly on the increasing *rate* of sea level rise since the mid-1800s. Over the past century the sea level in the Cape region has risen about 12 inches. There are also changes in the bedrock beneath the Cape Cod region. After rebounding in height after the glacier melted back from the region, the bedrock has begun to sink a few inches per century all along the Atlantic Coast. On the New England coast the rate of sea level rise in 2017 is about 3.5 inches per

decade on average, more than twice the average historic rate
during the twentieth century.

Glacial features in today's landscape

Virtually everything you see on the surface of today's Outer
Lands region reflects the work of the Wisconsinan Glacial
Episode glaciers, modified by about 15,000 years of weather-
ing and marine erosion.

Moraines

The great glacial moraines in today's Cape and Islands landscape are ironically a bit hard to see because they are so large.
The upland hills of Upper Cape Cod, the low, rolling hills of
Tisbury, Oak Bluffs, and Aquinnah on Martha's Vineyard, and
the highest points of Nantucket Island are all glacial moraine
areas. On the Upper Cape most of Route 6 between Sandwich
and Dennis runs on the crest of the Sandwich Moraine. The
long arc of the Elizabeth Islands and the hills just east of the
Buzzards Bay shoreline between Woods Hole and Sandwich are formed by the Buzzards Bay Moraine (see map, p.
25). Route 28 in Bourne and East Falmouth runs atop the
Buzzards Bay Moraine. On Long Island's eastern end the moraines are a bit more obvious: both the north and south forks
of Long Island are moraines. The central backbone hills of
Long Island are the Ronkonkoma Moraine, the southernmost
extent of the glacial ice 25,000 years ago (see illustration, pp.
20–21). The eastern end of the Ronkonkoma Moraine forms
Montauk Point. Perhaps the most visually obvious moraine
in the Outer Lands area is the Harbor Hill–Orient Point Moraine, whose eroded 50–100-foot cliffs form the north coast

Until about 8,000 years ago
the moraines and outwash plains
that formed the Cape and Islands
were *inland* hills well north of
the ancient Atlantic coastline.
The sea level gradually rose after
the Wisconsinan glacier melted
away, transforming these inland
hills into the Cape and Islands we
know today.

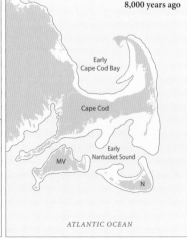

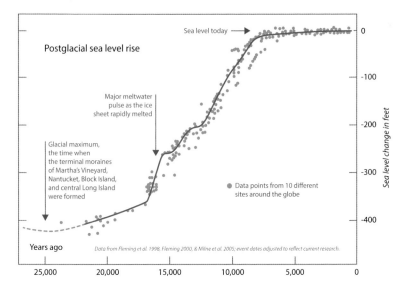

Postglacial sea level rise

Sea level today →

Major meltwater pulse as the ice sheet rapidly melted

Glacial maximum, the time when the terminal moraines of Martha's Vineyard, Nantucket, Block Island, and central Long Island were formed

● Data points from 10 different sites around the globe

Years ago

Data from Fleming et al. 1998, Fleming 2000, & Milne et al. 2005; event dates adjusted to reflect current research.

Sea level change in feet

After the glacial maximum 25,000 years ago, the sea level began to rise, at first gradually and then much more quickly as the glaciers melted. A series of meltwater pulses caused the sea level to rise quickly during several periods, and then the rate of sea level rise slowed about 8,000 years ago.

of eastern Long Island for almost 50 miles from Port Jefferson east to Orient Point.

Outwash plains

So much glacial drift was available after the glaciers melted that streams of meltwater carried huge amounts of smaller rocks and soil particles downstream, forming large plains of mixed rock and soil called outwash plains. Much of the Inner Cape south of the Route 6 highway and most of Nantucket and Martha's Vineyard consist of glacial outwash plains. Most of the lowland areas of Long Island are also outwash plains of mixed glacial drift that washed south from the Ronkonkoma and Roanoke Point Moraines as the glaciers melted.

Outwash plains are not uniform in composition. Meltwater running off the deteriorating glaciers sorted the glacial rubble by size, creating outwash plains that are a jumble of medium-sized rocks and mixed glacial clay and sand, occasionally dotted with boulders too large to be moved by meltwater streams. Near the small rivers that crossed the outwash plains there may be large areas of sand that washed out from the plains and onto nearby beaches. In many areas meltwater streams sorted the fine sediments in glacial rubble into distinct layers. The finest clay sediments tend to move the farthest, settling out of the water as glacial streams lost speed and energy crossing the broad outwash plains. Much of the clay sediment from the outwash plains has been carried out to sea over the centuries, but coastal currents also bring some of the clay particles back toward more sheltered waters where the accumulated clay helps build salt marshes and tidal flats.

Spring-sapping valleys

If you look at maps of the southern coastal areas of Cape Cod, Martha's Vineyard, Long Island, and the southwestern shoreline of Nantucket you see a similar pattern repeated many times: a pond or small bay along the shoreline, with many fingerlike branches extending inland to the north of the main pond or bay. This pattern is particularly obvious along the south shore of Martha's Vineyard: the Edgartown Great Pond, Ripley Cove, Long Cove, and Tisbury Great Pond all follow this same general shape. At first glance the origin of the pattern seems obvious: the branching shapes seem to be due to streams that flow south, feeding into the coastal ponds and bays. But when you study the areas you can see that there are no substantial rivers or streams—certainly none large enough to have eroded these small valleys. The valleys also tend to have relatively flat bottoms, unlike the typical "V" shape of valleys cut by surface streams and rivers. The narrow "river" mouths that branch north of the coastal ponds and bays could not have been cut by water flowing on the surface, so how did they form?

Geologists call these narrow valleys spring-sapping valleys, and many dozens of them appear along the south shores of the Outer Lands from Jamaica Bay, New York, all the way to Chatham on Cape Cod. Spring-sapping valleys formed as the glaciers melted and large meltwater streams crossed the

The eroded cliffs of the Harbor Hill recessional moraine at Wildwood State Park, Wading River, Long Island.

outwash plains. The soils of the Outer Lands were mostly bare glacial drift, without much vegetation cover to stabilize the sandy sediments. The sea level was several hundred feet below today's level, so the ocean coastline was many miles south of where it is today. As later groundwater from rain and snow drained from the uplands, it did so mostly as springs flowing through the very porous, sandy sediments, and the flowing water emerged above ground as springs and small streams well south of today's shoreline. These south-flowing streams "sapped" sand and clay from the glacial drift, and over many centuries the springs removed so much sediment that the sapping caused the ground around the springs to collapse into the shallow valleys we see today. As this sapping was creating the valleys, the sea level was rising, and the sea eventually flooded the south ends of the valleys, creating the branching pond pattern we see today. Later coastal erosion and longshore currents blocked the mouths of many coastal ponds with sandspits and barrier islands.

The formation of spring-sapping valleys. The left and right illustrations show the same area of Martha's Vineyard. At left, 10,000 years ago, the area was well north of the coastline, and small underground springs cut low channels in the ancient outwash plain. At right, the same area today with today's sea level shows that the old spring-sapping valleys have flooded, and the old south-flowing stream is now submerged beneath the sea.

Glacial boulders and erratics

A glacial erratic is a boulder moved by glacial ice that is different from the local bedrock underneath it. Because the bedrock of the Outer Lands is largely buried hundreds of

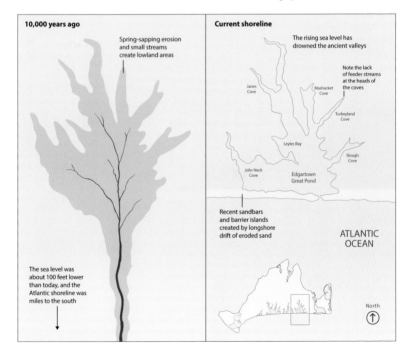

10,000 years ago

Spring-sapping erosion and small streams create lowland areas

The sea level was about 100 feet lower than today, and the Atlantic shoreline was miles to the south

Current shoreline

The rising sea level has drowned the ancient valleys

Note the lack of feeder streams at the heads of the coves

Janes Cove

Mashacket Cove

Turkeyland Cove

Leyles Bay

Slough Cove

Jobs Neck Cove

Edgartown Great Pond

Recent sandbars and barrier islands created by longshore drift of eroded sand

ATLANTIC OCEAN

North

feet below ground, most surface glacial boulders in the Cape and Islands region are glacial erratics, unrelated to the deeply buried bedrock. Millions of glacial boulders dot the landscape of the Outer Lands from Race Point through the length of Long Island. The largest and most famous of the glacial erratics in the Cape Cod area is Eastham's Doane Rock (also called Enos Rock), just east of the Salt Pond Visitor Center at Cape Cod National Seashore. Roughly 40 feet long and 25 feet wide, Doane Rock is so large that it looks like an outcropping of bedrock, but this giant boulder was actually torn loose by the South Channel Lobe as it passed over what is now a ridge of submerged volcanic bedrock in Massachusetts Bay. The huge rock later melted out of the glacier as the ice decayed and then was partially buried by surface erosion and settling. Geologists think that only about half of Doane Rock is currently exposed, making it by far the largest glacial erratic in the Cape Cod region.

Glacial boulders occur throughout the Outer Lands, but they are especially common on the Falmouth Peninsula and the base of the Upper Cape because the Buzzards Bay Lobe ran over the nearby exposed bedrock of the eastern Massachusetts mainland, and the boulders there had a shorter route to travel within the glacial ice before being deposited on the peninsula and the nearby Elizabeth Islands.

Kettle ponds and kettle holes

Cape Cod's outwash plains were at times deep under ice, and as the ice sheet melted it left huge remnant chunks of glacial ice partially buried in the glacial rubble and outwash sediments. The glacial ice eventually melted, leaving large, deep craters in the outwash plains. Ponds formed in craters deep enough to intersect with the local water table. These small bodies of water are called kettle ponds because they are often round or nearly so and are very steep-sided and deep compared to ponds formed in low areas by streams. Not every remnant chunk of glacial ice formed a kettle pond. Kettle holes are dry (or seasonally wet) depressions formed like kettle ponds but with their bottom above the water table. Kettle holes are sometimes home to wild cranberry bogs, and some commercial cranberry bogs take advantage of natural kettle depressions in the Cape Cod landscape.

More than 500 kettle ponds are scattered across the outwash plains of Cape Cod, and kettle ponds and kettle holes occur throughout the Outer Lands. Lake Ronkonkoma, Long Island's largest freshwater lake, is a kettle pond. The outwash plains of Martha's Vineyard and Nantucket were never completely covered with ice because they lie just south of the

Doane Rock is the Cape's largest glacial boulder. It lies just east of Cape Cod National Seashore's Salt Pond Visitor Center in Eastham.

southern edge of the main ice sheet, and thus there are fewer kettle ponds on the islands, and their kettle ponds are much smaller than those on the Cape. The kettle ponds of the Outer Lands are typically fed by groundwater and local runoff from rain and snow, and are not usually supplied or drained by substantial surface streams. The rounded shorelines of kettle ponds are not due to their icy origins but are the effect of centuries of wind and surface erosion evenly distributing the sediments around the pond edges.

Some kettle ponds near the coast have been invaded by seawater. Chatham's Stage Harbor, Salt Pond next to the Cape Cod National Seashore Visitor Center in Eastham, and Quissett Harbor in Woods Hole are examples of former kettle ponds that are now flooded by tidal water. On Long Island, Lake Montauk, Napeague Harbor, and Three Mile Harbor are all large kettle ponds that were later invaded by seawater. The interior of eastern Long Island is also dotted with small kettle ponds, and it is possible that some of the small bays and inlets along Peconic Bay and Gardiners Bay are the remains of former kettle ponds now flooded by the sea.

Kettle ponds can be quite substantial lakes. This is Gull Pond in Wellfleet.

Marine erosion and deposition in the Outer Lands

Since the melting of the Wisconsinan ice sheet the modern coasts of the Outer Lands have been shaped chiefly by marine erosion—the strong, steady beat of waves on beaches, occasionally punctuated by hurricanes and nor'easter storms. Once the ice sheet left the Outer Lands area the Cape and Islands region was a large inland area of low, rocky hills and muddy outwash plains, dotted with huge melting chunks of ice that later became kettle ponds. The sea level rose as the great ice sheets melted away. The land that had been covered with ice also rose: the tremendous weight of the glacial ice that had formerly depressed the earth's crust was lifted, and the land rebounded to its former height. The flooding of the

Cape region was thus a gradual process that took thousands of years (see illustrations, p. 31). By about 6,000 years ago the sea level reached a point roughly 50 feet below today's sea level, and the ocean waves began eroding and shaping the Cape and Islands area. By about 3,500 years ago the waters were about 20 feet below present sea level, and the Cape and Islands would have been recognizable as a great peninsula and two islands, but much larger than they are today, with a much rougher coastline of bays and newly formed headlands. The past 3,500 years of marine erosion and deposition of sand created the Outer Lands shorelines we see today.

The action of waves

Ocean waves form as wind moves over the surface of the sea, and the friction drag on the water surface forms ripples that consolidate into larger waves. Strong ocean winds then further drive the waves. The process is largely a matter of energy transfer: the sun warms the atmosphere and the solar energy creates winds. The winds move over the ocean, transferring some of their energy into the surface waters. On average 6,000–8,000 waves per day hit the exposed ocean beaches of the Outer Lands. As waves reach land, the impact

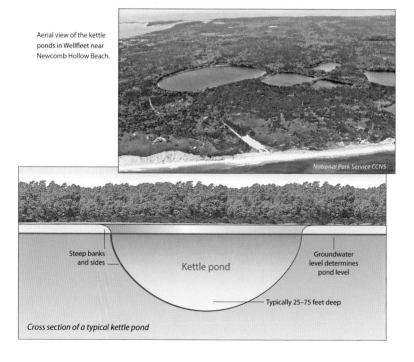

Aerial view of the kettle ponds in Wellfleet near Newcomb Hollow Beach.

National Park Service CCNS

Steep banks and sides

Kettle pond

Groundwater level determines pond level

Typically 25–75 feet deep

Cross section of a typical kettle pond

of their breaking provides enough momentum over time to move mountains of sediment from the coast and into coastal currents. Although the glaciers contributed the earthy substance of the Outer Lands, most of what you see along the modern shoreline reflects the work of wave energy eroding and sculpting that raw glacial drift into the Cape and Islands that we see today.

Cape Cod, Martha's Vineyard, and Nantucket were once much larger than they are today, and this shrinkage is mostly due to the eroding power of ocean waves, particularly during winter storms and late summer hurricanes. A fair-sized ocean wave can break onto the beach with a force as great as two tons per square foot. Over the past 6,000 years ocean waves have both removed large amounts of glacial sediments and redeposited about 25 percent of those sediments to the coastline in the form of sandy beaches and spits, as well as to the salt marshes and mud flats of the more protected bays and inlets.

Through the processes of erosion and deposition, waves and longshore currents smooth the shoreline. Headlands are worn down by the waves, and the material removed from the headlands tends to be spread along the surrounding coast by longshore currents that run parallel to the shoreline. The net effect of wave action is to smooth out coastlines that were once quite lumpy, with irregular headlands and deep bays.

Outer Cape Cod was once much wider than it is today (see illustrations, p. 31). The exact details of the ancient Cape coastline can only be guessed at, but as the rising sea level reached the regions near the present Cape, the shorelines probably

In the Outer Lands, soft earthen headlands exposed to ocean waves erode at an average rate of about three feet per year, but averages are misleading: most major erosion of cliffs such as these at Aquinnah on Martha's Vineyard happens suddenly during winter storms, when huge chunks of water-sodden clay avalanches down the cliffs. In such incidents a cliff face may lose more earth in a few moments than in many years of normal weathering.

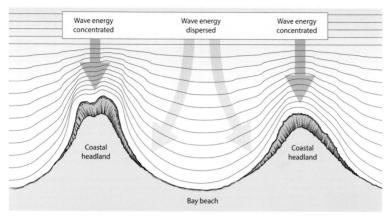

looked quite complex, with many offshore islands, bays, and smaller water channels. Wave action smoothed many of the headlands and created large sandspits like Sandy Neck and Monomoy Island. In more sheltered areas like Cape Cod Bay the silt and clay settled out of the water and became the foundation for salt marshes. Today Cape Cod, Martha's Vineyard, and Nantucket collectively lose about five acres per year to erosion, although some of that lost material is deposited by longshore currents. Eastern Long Island and Block Island lose a comparable amount through the same erosion processes.

Erosion of headlands by waves. Wave action tends to focus energy on headlands that project from the coast, eroding them much faster than the relatively sheltered bays in between. As the wave action removes projecting headlands, the coastline becomes smoother.

Marine scarps or ocean cliffs
In the most exposed areas of the Cape shoreline the waves have cut away the glacial sediments so aggressively that there

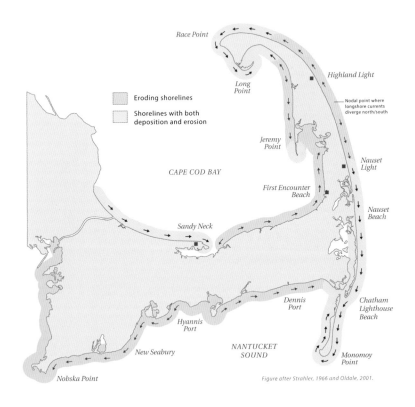

Race Point

Long
Point

Highland Light

Eroding shorelines

Shorelines with both
deposition and erosion

Nodal point where
longshore currents
diverge north/south

Jeremy
Point

Nauset
Light

CAPE COD BAY

First Encounter
Beach

Nauset
Beach

Sandy Neck

Dennis
Port

Chatham
Lighthouse
Beach

Hyannis
Port

NANTUCKET
SOUND

New Seabury

Monomoy
Point

Nobska Point

Figure after Strahler, 1966 and Oldale, 2001.

The general direction and erosive effect of longshore currents on the shores of Cape Cod. The currents are driven by general seasonal wind patterns and by larger regional currents in the Gulf of Maine, Cape Cod Bay, and Nantucket Sound. Only about 25 percent of material eroded from beaches and cliffs is redeposited to form new or larger beaches, sandspits, and barrier islands.

is no gentle coastal slope down to the beach: we just have the high, sheer cliffs or marine scarps (as geologists call them) that run 18 miles north along the Cape's ocean shoreline from Eastham to High Head in Truro. In areas where the glacial sediments are soft, the scarp face angles downward about 30–35 degrees, about all the slope angle that the soft glacial sediments will support before they collapse. In firmer clay sediments the scarps form towering 150-foot near-vertical cliffs over the beach in such areas as the Cape Highlands near Highland Light (Cape Cod Light) in Truro. Large sea cliffs are found throughout the Outer Lands, and include Sankaty Head on Nantucket, the famous Gay Head (named for the bright colors of the sediments) in Aquinnah on Martha's Vineyard, Mohegan Bluffs at the southern end of Block Island, and the almost 50 miles of 75–100-foot sea cliffs that form the north shore of eastern Long Island.

The marine scarps (cliffs) above the Great Beach between Eastham and High Head in Truro lose an average of two to three feet a year due to erosion, mostly in winter storms.

Sea cliffs are the glory of Cape Cod, offering sweeping views of both the ocean and the rolling highland moors and pine forests. Building near such magnificent views has been an attraction since colonial times, but the rate of erosion of the marine scarps is so rapid that most structures built near the cliffs that are more than 60 years old have either already vanished over the edge in storms or been moved back from their original locations at least once. Cliff erosion on the Cape has averaged three feet per year in historic times, but an "average rate" for cliff erosion is a misleading concept. Most of the erosion happens during winter storms, when huge sections of cliff can collapse onto the beaches below and many feet of cliff face are lost in moments. In May 2016 a large section of the ocean cliffs above Coast Guard Beach in Truro suddenly gave way in a classic bank slump: a huge segment of waterlogged clay and sand sediments on the cliff face dropped without warning onto the beach at the base of the cliff. Pay attention to the many warnings you see to stay well away from the edge of sea cliffs, as they can give way without warning even in hot, dry weather.

Marine scarps (cliffs) at Lecount Hollow Beach, Wellfleet. The cliffs above Lecount Hollow Beach contain a lot of clay, which will maintain a steeper angle of slope than looser, sandier glacial deposits. Most large landslides happen during winter storms, but please pay attention to the beach signs warning against climbing the cliffs: landslides can happen at any time of year.

The 150-foot marine scarps of the Mohegan Bluffs on Block Island, Rhode Island.

A visit to the Cape's famous Highland Light will give you insights into the reality of erosion and shoreline change, as you can see both the light's current location and the marker for the older 1857 location of the light. The light's original 1797 location has long since vanished due to the eroding cliffs.

The history of the Cape's Highland Light is instructive. Highland Light was the first lighthouse constructed on Cape Cod. The original wooden structure lasted from 1797 until 1857, when it was deemed unsafe and demolished because erosion had cut away so much land that the light was in danger of falling over the cliff edge. The current stone structure was built in 1857 about 500 feet back from the cliff edge. By 1990 the setback from the cliff had eroded to about 125 feet, and the lighthouse and keeper's house were moved back another 450 feet to the present location. The cliff edge near the lighthouse erodes at a rate of about five feet per year, so the light will have to be moved again in about 50 years to save it from destruction, assuming that the current rate of erosion stays the same.

Longshore currents and beach drifting

Sand particles don't just move straight up the beach as waves swash in and straight down the beach as the backwash slides away again. Because of the effects of wind and ocean currents, wave sets rarely meet the beach in perfectly parallel lines, and they almost always come in at some angle to the line of the beach. The constant angle of waves to the beach has two major effects: the angled waves create a current just offshore of the beach, called a longshore current. The angled waves in the foreshore zone of the beach also move sand particles along the beach in the direction of the longshore current. Typical longshore currents flow at two to three miles per hour along the shore. If you've ever been swimming in the ocean surf for a while and ended up coming ashore well down the beach from your towel and umbrella, blame the longshore current. Although the amount of sand moved by one wave is modest, the accumulated action of thousands of waves a day is enough to move many hundreds of tons of sand down the beach over a year. The old saying that "it's a new beach every year" is true: if you visit the Cape's Marconi Beach this July, most of the sand you will be standing on was transported about a mile down the beach from the north by the storms of winter and the longshore currents that move sand all year.

Sandbars and sandspits

Where the coastline bends sharply and the water deepens, such as at the mouth of a bay or the end of a peninsula, the longshore current slows sharply and the sand particles it carries settle out, gradually forming a sandbar. Over time this sandbar accumulates more sand and becomes a sandspit above the high tide level. Each year, more sand accumulates at the free end of the sandspit, further lengthening the spit. Once the sand is permanently above the high tide level, vegetation moves in and helps accumulate yet more wind-driven

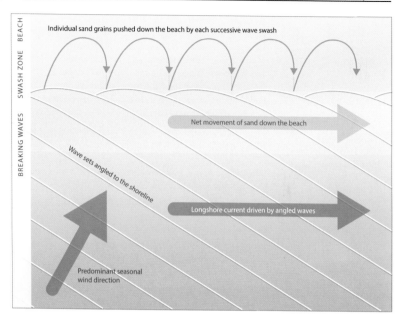

Individual sand grains pushed down the beach by each successive wave swash

Net movement of sand down the beach

Wave sets angled to the shoreline

Longshore current driven by angled waves

Predominant seasonal wind direction

BEACH

SWASH ZONE

BREAKING WAVES

sand grains that are trapped and held by the leaves and stems, forming sand dunes on the spit. Nauset Beach in Eastham, Monomoy Island south of Chatham, Sandy Neck in Barnstable, and the whole northern tip of Cape Cod northwest of High Head in Truro are all examples of sandspits formed by longshore drift.

Beach sand movement and the longshore current.

When large sandspits extend from eroding headlands and begin to enclose bays between the sandspit and the mainland, they are known as barriers. Overwash from storms often cut off sandspits from their original headland sources, and the sandspits become true islands, but the process of erosion and sand deposition continues.

On the south coast of Long Island there are large barrier islands and substantial sandspits, reflecting the greater supply of sand in an area where the erosion of Long Island brings more sand to the coastline. Fire Island and the long sandspits that enclose Shinnecock Bay, Moriches Bay, and the Great South Bay are all derived from the coast erosion and deposition processes that form sandspits from New York's Montauk Point all the way down the Atlantic Coast to the south tip of Miami Beach.

The Nauset Spit and Nauset Beach series of sandspits runs from Eastham south to Chatham, and then Monomoy Island

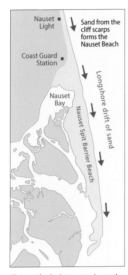

The sandspit that extends south of Eastham's Coast Guard Beach is a classic coastal sandspit, supplied by erosion from the oceanside cliffs of Eastham and Wellfleet.

carries the same sandspit system 10 miles farther south. The Nauset Spit is a small southern counterpart to the northern Cape's Provincelands. The predominant longshore currents south of Truro run south along the shore, and Nauset Beach and Monomoy Island are built largely from sand eroded from the marine scarps of the Cape from the north border of Well-fleet south to the southern end of the scarps at Nauset Light Beach and Coast Guard Beach.

The changing configurations of the Nauset Spit and Monomoy Island show how tenuous life can be when you rely on land-forms made of sand. The shifting barrier islands and sandspits off Chatham have been a constant challenge to mariners since the town was founded in the late 1600s.

The northern tip of Cape Cod, called the Provincelands or sometimes the Province Lands, is essentially one giant sandspit, formed by a combination of material eroded from the marine scarps of the Outer Cape and the action of the north-flowing longshore currents off the eastern coast of the Outer Cape.

The formation of the Provincelands

The history of the Provincelands begins about 8,000 years ago, when the last remnants of the ancient Georges Island were finally submerged by the rising sea level and became Georges Bank. Once Georges Island no longer sheltered the Outer Cape from the force of Atlantic storms and ocean waves from the southeast, the erosion of the eastern Cape shoreline accelerated, and the dominant longshore current pattern on the Cape's ocean coast shifted north in the Truro area. The rising sea level also accelerated the overall rate of erosion on the Outer Cape, which was once about three miles wider than it is today. These two factors brought millions of tons of new

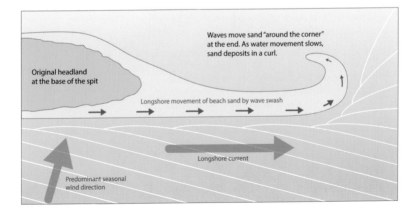

sand and sediment into the north-flowing longshore current patterns every year, and a new sandspit began to grow north from the marine scarps of High Head in Truro, which 4,000 years ago was the northern tip of Cape Cod.

About 3,500 years ago the sandspit that became the Provincelands reached the area of what is now downtown Provincetown, and over the next thousand years this initial sand ridge formed the first growth ring of the Provincelands area. This early dune line is still intact, and Bradford Street in Provincetown and the Pilgrim Monument lie along the crest of the original sand ridge. Later sandbars formed in successive growth rings north of the original sandbar and gradually became incorporated into the mainland beach and dunes of the early Provincelands (see illustrations, p. 48). Each sandspit built out from the High Head area as a hooked spit, and each successive hook enlarged the width of the Provincelands peninsula. Each new sandspit also hooked southward at its western tip, creating the widening expanse of the fist of the Cape's arm west of the narrow wrist that connects the Provincelands to High Head in Truro.

The Peaked Hill Bar that parallels the modern northeastern shoreline of the Cape seems to be forming yet another northward extension of the Provincelands sandspit. The bar has been present in similar form and location since colonial

Fire Island, New York, shown here at Robert Moses State Park, is a long, east-west complex of barrier islands and sandspits just off the southern coast of Long Island.

The Provincelands at the northern tip of Cape Cod.

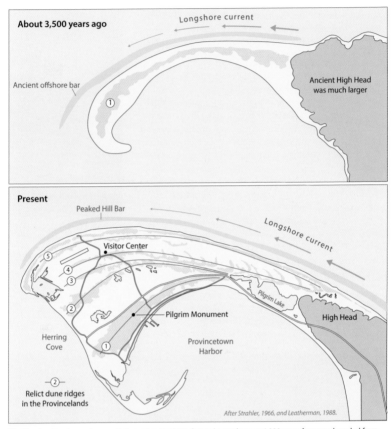

About 3,500 years ago

Longshore current

Ancient offshore bar

①

Ancient High Head was much larger

Present

Peaked Hill Bar

Longshore current

Visitor Center

⑤
④
③
②
①

Pilgrim Lake

Pilgrim Monument

High Head

Herring Cove

Provincetown Harbor

—②—
Relict dune ridges in the Provincelands

After Strahler, 1966, and Leatherman, 1988.

The Provincelands area is essentially a giant sandspit, formed over the past 6,000 years from sand eroded from the oceanside cliffs of the Cape. The Provincelands have been built in a succession of sand dune ridges that can still be seen in the landscape today. Peaked Hill Bar may one day become the next ridge in the series.

times. The British frigate HMS *Somerset* went aground on the Peaked Hill Bar near High Head in Truro during a storm in 1778, and to this day the shifting beach sands near High Head periodically expose pieces of the wreck. The slow rate of change in the Peaked Hill Bar suggests that it may take as much as 1,000 years for an offshore bar to convert to a new extension of the Provincelands. The building rate of new material is also slowed by the depth of the waters off Race Point and Long Point, which are as much as 200 feet deep within a few hundred yards of shore. In spite of these challenges, Race Point and the nearby Long Point continue to shift position and extend, sometimes more than 100 feet in a single year.

The Pilgrim Monument in Provincetown sits high on the first relict dune ridge of the Provincelands, as does the length of Bradford Street (Route 6A).

The Beech Forest in the Provincelands gives us a tantalizing glimpse of the lush forests that once covered these dunes before early European settlers cut the trees down for building material and firewood.

Sandy Neck in Barnstable is the other major sandspit and dune area on Cape Cod and is in many ways more accessible to hikers than the much larger but more remote dune areas of the Provincelands. Sandy Neck has four cross-dune trails that will show you some of the Cape's most beautiful dune ridges, rising above the huge salt marshes of Barnstable Harbor.

A Humpback Whale feeds in the rich waters of Stellwagen Bank just north of Cape Cod.

WEATHER AND WATER

The green waves of Race Point Beach are rich in plankton and other marine life.

The Outer Lands are a series of islands and peninsulas surrounded by the Atlantic Ocean, the Gulf of Maine, and local bays and sounds contiguous with ocean waters. The weather of the Outer Lands and the ocean life off its shores are influenced and enriched by the clash and mixing of nearby warm and cold ocean currents.

On the Cape and Islands the weather generally comes from west to east, driven by the predominantly westerly band of winds at this latitude in the northern hemisphere. The great exceptions to this general pattern are major low-pressure storms that arrive from the south and southwest: hurricanes in the late summer and early fall, and nor'easter storms that mostly occur in the colder months of the year.

As landforms surrounded by large bodies of water, the Outer Lands are always windier than inland areas. The predominant wind patterns in the Outer Lands are strong, cold northwesterly winds* in the winter and early spring, and gentler, warm southwesterly winds in the late spring through early fall. In the sandy dune and beach areas and exposed coasts of the Outer Lands the wind plays a large role in shaping the land, particularly the strong northwesterly winds of winter. The shape and character of the sand hills in major dune areas like the Provincelands at the Cape's northern tip and in Barnstable on the Sandy Neck peninsula are largely created and shaped by northwesterly winter winds. In late spring, summer, and early fall the predominant southwesterly winds help bring the warmth and moisture of the southeastern United States and Gulf Stream northward into the Outer Lands.

*Winds are named by the direction they come from. Northwesterly winds come from the northwest.

The average temperature and rainfall profiles for Hyannis, Cape Cod, are typical of the region.

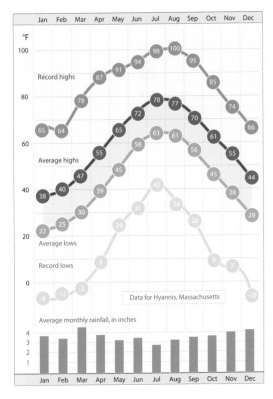

The ocean has a strong influence on temperatures in the Outer Lands. In fall and early winter the relative warmth of the ocean moderates Cape and Islands temperatures. In spring leaves and flowers appear weeks later along the coasts than in inland areas of southern New England, because the cold ocean keeps average coastal temperatures about 10 degrees Fahrenheit lower than inland temperatures.

A transition and border region

The large peninsula of Cape Cod and nearby islands are a climactic transition zone between the relatively mild temperatures, warmer waters, and low, sandy coasts of the mid-Atlantic coastal region of the United States and the cold water and generally rocky coasts of the Gulf of Maine. In late summer the waters of Nantucket Sound can be as warm as 73 degrees Fahrenheit, whereas the water in the middle of Cape Cod Bay rarely exceeds 65 degrees Fahrenheit. The Outer Lands form the northern limit for many species of plants and animals, particularly marine and estuary animals like the Blue

Crab, fiddler crabs, and the Bay Scallop. On land Cape Cod is the northern limit for the Atlantic coastal pine barrens forests more typical of Long Island, New Jersey, and the central Atlantic Coast, and the Cape is almost the northernmost location for the American Holly. The Cape is also the southern limit for cold-adapted plants like the cranberry, bearberries, and many birch species, which are more typical of the Maine coastline. The cranberries and birches are probably remnants of the northern forest and tundra plant communities that developed as the Wisconsinan glaciers retreated about 20,000 years ago.

The southeastern New England and Long Island coastal regions are affected by two major ocean currents: the cold Labrador Current that flows south from the Arctic Ocean along the Atlantic Coast and the warm Gulf Stream that flows northeastward several hundred miles south of Long Island and Cape Cod.

The Labrador Current
The Labrador Current originates in Baffin Bay off Greenland and flows south along the Labrador and Newfoundland coasts and over and around the Grand Banks area. Branches of the Labrador Current chill the waters of the Gulf of Maine and flow south along the continental shelf past the Cape and Islands as far south as Cape Hatteras. In the Outer Lands these cold currents form biologically rich frontal zones where they

The collision and interaction of the cold waters from the Labrador Current with the water from the Gulf Stream to the south are the major oceanic influences on local weather around the Cape. The contrast in water temperatures can generate thick fog banks at any time of year, particularly over Chatham, Nantucket, and Georges Bank.

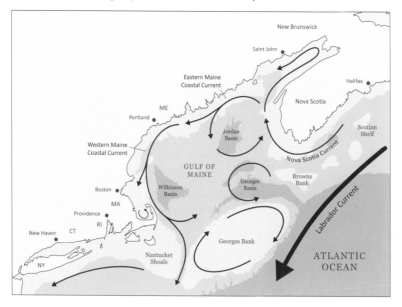

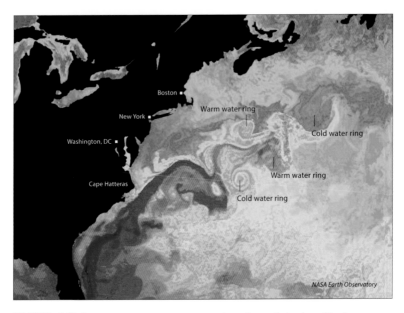

NASA Earth Observatory

This NOAA satellite image of the Gulf Stream codes water temperatures from blue (cold) to very warm (dark red). The rings are warm or cold gyres, huge circular areas that cycle off the main Gulf Stream and often bring tropical fish and birds close to coasts of the Cape and Islands.

meet warmer water from the south, in places like Georges Bank and the continental shelf edges south of southern New England. In frontal zones (see next page) cold currents from the north helps bring nutrient-rich water close to the ocean surface, and warm water from the south promotes growth of phytoplankton (microscopic marine plants) and zooplankton (microscopic marine animals). Plankton is the base of the ocean food chain, and in plankton-rich waters fish and ma-rine mammals from both cold-water and warm-water regions meet to feed.

The Gulf Stream

The Gulf Stream is a giant heat engine, warming the land and waters of the New England coastline even though the nearest edge of the stream is usually more than 200 miles southeast of the shores of the Outer Lands (see illustration, above). Averaging about 60 miles in width, the Gulf Stream is well over 1,000 feet deep as it flows northeast off Cape Hatteras. Traveling at an average speed of 5.6 miles per hour, the stream moves about 4 billion cubic feet of water per second, more water than all the world's rivers combined.

The Gulf Stream has multiple direct effects on the marine ecology and weather of the region, bringing tropical and mid-Atlantic coastal fish species to the New England area, particularly in late summer and early fall. As a river of water within the larger ocean, the current meanders as it runs south of New England, and large bends in the central flow of the Gulf Stream often curve north and approach the continental shelf off the Outer Lands. The meander bends frequently break away from the main current of the Gulf Stream, forming giant circles of warm water north of the stream. These warm-core rings from the Gulf Stream typically drift toward the northwest and the New England coast, and within the ring the water rotates in a counterclockwise direction. Gulf Stream water within the warm-core ring brings warm-water plankton and larger tropical animals close to the coasts of the Cape and Islands. Warm-core rings are often 100 or more miles in diameter, extend from the ocean surface to depths of 1,000 feet or more, and can persist for as long as two years.

The collision of currents off the Outer Lands

Frontal zones where the cold water of the Labrador Current and Gulf of Maine meets warm offshoots of the Gulf Stream help create the rich and diverse marine ecosystems that surround the Outer Lands. Off Cape Cod the swift tidal currents that sweep over the shallow Georges and Stellwagen Banks and the Nantucket Shoals mix with warm and cold ocean currents to create some of the richest and most varied ocean life in the Western Atlantic Ocean. In NASA's panoramic view of this ocean region (see illustration, pp. 58–59), the deep green swirls of phytoplankton over the Nantucket Shoals, Georges Bank, and the frontal mixing zone at the continental shelf edge just south of Georges Bank are where the largest concen-

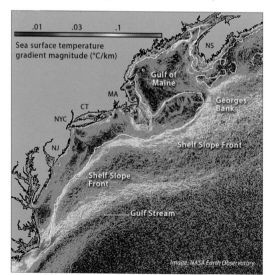

.01   .03   .1

Sea surface temperature
gradient magnitude (°C/km)

NS

Gulf of
Maine

MA

Georges
Bank

CT

NYC

NJ

Shelf Slope Front

Shelf Slope
Front

Gulf Stream

Image: NASA Earth Observatory

This sea surface temperature gradient map highlights areas of the Western Atlantic where the cold waters of the Labrador Current collide with offshoots of the warm Gulf Stream. These frontal collision zones (deep red and yellow in the image) are particularly rich in marine life of all kinds. The Shelf Slope Front is a zone of cold water that derives from the Labrador Current and follows the edge of the continental shelf, where it drops off into very deep ocean waters.

trations of predatory fish species, marine turtles, whales, and dolphins congregate. The more inshore waters of Nantucket Sound, Cape Cod Bay, Stellwagen Bank, and Block Island Sound are also rich in phytoplankton, marine mammals, and fish species.

Besides rich fishing and great whale watching, the collision of warm and cold water off the Outer Lands creates another characteristic New England coastal feature: dense fog, particularly in summer and fall. When warm, moist air from the Gulf Stream is swept north by the prevailing southwesterly winds of summer, it meets the cold waters of the Labrador Current around the ocean-facing shores of the Cape and Islands, and the moisture in the warm air begins to condense as fog. Fog at ground level normally requires a relative humidity of 100 percent, but over ocean waters fog droplets can form around tiny salt crystals in the air, and dense marine fogs can form at humidities as low as 97 percent.

Springtime in the northwestern Atlantic Ocean. The lighter green patches and swirls are dense swarms of phytoplankton, indicating regions where there are significant nutrients near the surface and other water conditions that promote blooms. Note how rich the shallows of the Nantucket Shoals and Georges Bank are compared to the deeper ocean areas. The Labrador and Gulf Stream currents meet in this area, forming the especially productive red areas visible on the southern edge of Georges Bank and the large green swirls of phytoplankton at lower right.

Nor'easters

Nor'easters are storms that typically originate in the Gulf of Mexico as warm, moist, low-pressure systems that are then steered northeast across the south-central United States by the prevailing jet stream winds, eventually tracking north-northeast paralleling the Atlantic Coast. Nor'easters are particularly likely when a large high-pressure system sits over the Bahamas area, as this forces the storms off their usual eastward track and toward the northeast. Although a nor'easter can appear at any time of year, these storms are more common in the cold months between October and March, when they can bring devastating winds, large coastal storm surges, and

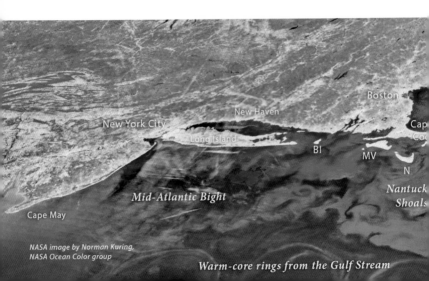

NASA image by Norman Kuring, NASA Ocean Color group

Warm-core rings from the Gulf Stream

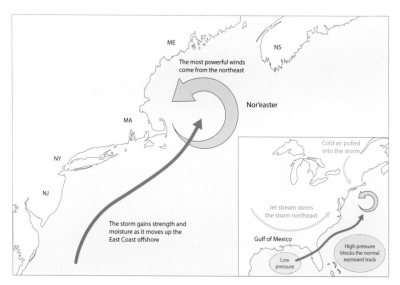

blizzard conditions to coastal regions of the United States and the Canadian Maritime Provinces. A large nor'easter can be as destructive as any hurricane and can cause major erosion in the soft earthen coastlines of the Outer Lands.

As a low-pressure system, a nor'easter circulates in a counter-clockwise motion and can be pictured as a circular clock face for points of reference (see illustration, above). As the storm tracks along the coast, the counterclockwise winds circulating from about 5 o'clock to 10 o'clock blow freely across the

A nor'easter moving up the Atlantic Coast, showing a typical storm track and counterclockwise wind circulation.

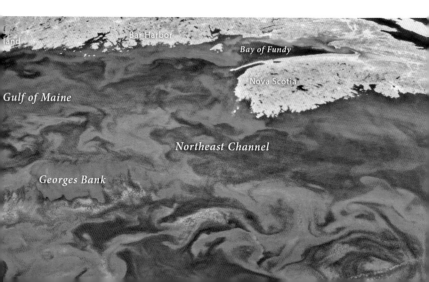

ocean and pick up speed and moisture. Observers along the coast will experience high winds coming onto shore from the northeast direction—hence the name "nor'easter." In a powerful nor'easter the winds off the ocean can pile up large waves and hurricane-like storm surges of up to 20 feet on ocean shores, flooding coastal communities and causing shoreline erosion. Winter nor'easters also bring snow. The largest recorded blizzards along the East Coast were nor'easters, such as the famous blizzards of 1978 and 1996, which both dumped two to three feet of snow along large sections of the East Coast in just a few hours. Winter storm Nemo in February 2013 was a classic blizzard-generating nor'easter, where a huge, moist, low-pressure system traveled north along the Eastern Seaboard and met another low-pressure system coming east out of the central United States, triggering a large blizzard that dumped record amounts of snowfall over New England. Severe nor'easters don't just bring snow. The winds and waves from nor'easters accelerate erosion along the coast and can make long-lasting changes, particularly in barrier islands, sandspits, and coastal cliffs.

Hurricanes

Hurricanes are tropical cyclones that are typically born as low-pressure systems off the west coast of Africa that then track westward across the Atlantic, gaining heat energy and moisture from the tropical midocean and arriving on our side of the Atlantic with storm-force winds and heavy seas. As they approach the Atlantic Coast the north-tracking hurricanes gain additional energy from the hot Florida Current

Data from the US National Oceanic and Atmospheric Administration (NOAA) records of the tracks of 69 tropical storms and hurricanes that have passed over the region since 1900.

at the base of the Gulf Stream, and almost every year these tropical storms hit parts of the East Coast. In the Outer Lands region these summer and fall storms are the warm-weather counterparts to winter nor'easters and are a major factor in changing and eroding local coastlines. A single major hurricane can cut more earth from the coast than a decade's worth of slow and steady erosion from the usual weather and waves. In the worst hurricanes the wind pattern is very similar to a nor'easter: heavy winds arrive on the coast from the northeast due to the counterclockwise circulation in these low-pressure storms.

Although direct hits from major hurricanes are unusual on the relatively cool coastlines of New England and Long Island, over the past 110 years, 69 tropical storms and hurricanes have tracked over southern New England (see illustration, p. 60).

Sea level rise and climate change

Waves, winds, and storms have the most immediate visible effect on the shape and shorelines of southern New England, and the sea has been rising steadily since the end of the Wisconsinan Glacial Episode about 25,000 years ago. But much of the recent concern has centered not just on the higher water levels but also on the accelerating rate of sea level rise.

We think of the shorelines we have known all our lives as permanent, but that is only because of our limited time perspective. In colonial times the sea level along the East Coast was more than two feet lower than it is today, which

Two hundred years ago these fields were planted with corn or wheat. Since then the sea level has risen more than two feet, and the stone walls now stand in high salt marsh. Barn Island, Stonington, Connecticut.

helps explain why you sometimes see colonial-era stone farm walls in today's New England salt marshes. When the walls were built in the mid- to late 1700s the land was arable and was probably planted with corn or hay, but in the intervening centuries the sea rose and the land around the stone walls became salt marsh.

Warmer global temperatures push up the sea level in two ways: through the melting of land-based polar ice caps and glaciers, adding more water to the ocean, and by thermal expansion of ocean water, because warm water occupies more volume than cold water.

Rising sea levels are no longer an ancient geologic curiosity. Global climate change—along with the sharp acceleration of the long-term trend of rising seas—means that rising sea levels will have a significant and visible effect on shoreline residents and visitors over the coming decades. The sea level is now rising by more than an eighth of an inch per year, and in the Outer Lands the sea level has risen about seven inches since 1960. Experts on climate change predict that the sea level along the Atlantic Coast will rise by approximately four to eight feet by the year 2100, and this rise will have profound impacts on salt marshes, beaches, coastal homes, and other coastal business and transportation infrastructure. Another factor is increasing sea level rise along the East Coast: long-term regional geologic trends are causing the land in coastal New England to sink by about a sixteenth to an eighth of an inch per year, worsening the apparent effects of sea level rise.

Geologists estimate that low-lying Nantucket has about 700 years before wave erosion and sea level rise turn the island into the northernmost part of the Nantucket Shoals. Pictured: Brant Point and its lighthouse on Nantucket Harbor.

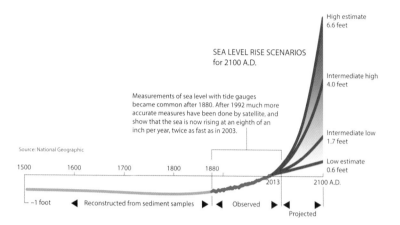

SEA LEVEL RISE SCENARIOS
for 2100 A.D.

High estimate
6.6 feet

Intermediate high
4.0 feet

Measurements of sea level with tide gauges
became common after 1880. After 1992 much more
accurate measures have been done by satellite, and
show that the sea is now rising at an eighth of an
inch per year, twice as fast as in 2003.

Intermediate low
1.7 feet

Source: National Geographic

Low estimate
0.6 feet

1500   1600   1700   1800   1880   2013   2100 A.D.

−1 foot   ◄ Reconstructed from sediment samples ►   ◄ Observed ►   ◄ ► Projected

Sea level rise will profoundly change the Outer Lands within
the lifetime of most readers of this book. Based on estimates
of a four-foot rise sea level rise over the next century, about
5,700 homes, 22,000 acres of land, and 120 miles of road on
Cape Cod will be permanently flooded. That four-foot rise,
moreover, is now the minimum prediction. Most estimates
of sea level rise predict up to a six-to-eight-foot rise over
the next century. The estimated value of the affected Cape
infrastructure is $4.9 billion in today's dollars. In this scenario
large sections of such low-lying wild environments as Monomoy Island, much of Sandy Neck, the Nauset and Chatham
sandspits, and a large portion of Provincelands around the

Martin Lehotkay

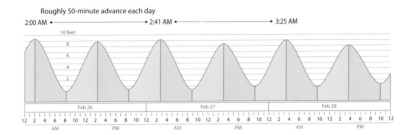

Roughly 50-minute advance each day

2:00 AM ← → 2:41 AM ← → 3:25 AM

airport and Pilgrim Lake will flood and erode into large sandbanks and tidal flats.

Tides and coastal zones

Apart from the moment-by-moment action of waves on the shoreline, the tides most strongly define the movements of water along our coasts. The tidal movements of ocean water convey life and nutrients to such shoreline environments as salt marshes, beaches, and tidal flats. Tidal water movements are also important farther offshore, where tidal currents sweep nutrient-rich water over such shallows as Georges and Stellwagen Banks, the Nantucket Shoals, Cape Cod Bay, and the Nantucket and Block Island Sounds.

Ocean tides are caused primarily by the gravitational pull of the moon and to a lesser extent by that of the sun. The gravitational effects of the moon and sun on earth's waters are complex, and the shape and depth of local landforms and the sea bottom further influence the depth and timing of tides.

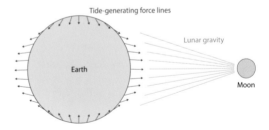

Tide-generating force lines

Lunar gravity

Earth

Moon

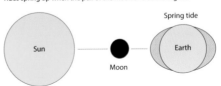

Tides spring up when the pull of the moon and sun is aligned

Spring tide

Sun

Moon

Earth

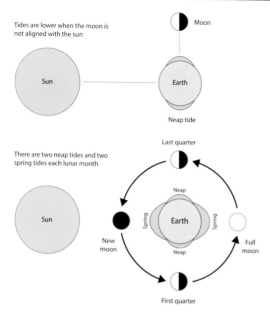

Tides are lower when the moon is not aligned with the sun

Moon

Sun

Earth

Neap tide

There are two neap tides and two spring tides each lunar month

Last quarter

Sun

New moon

Neap

Spring

Earth

Spring

Neap

Full moon

First quarter

The moon's gravity and position relative to the earth are the strongest influences on the height and timing of ocean tides. As the moon rotates around the earth once every 27.3 days, its gravitational pull creates a slight bulge in the ocean surface closest to the moon. The lunar day—the time it takes the moon to rotate once around the earth—is 50 minutes longer than the solar day, and so the tide cycle advances 50 minutes every day according to our clocks and calendars. These lunar or astronomical tides are also semidiurnal, rising and falling twice during each 24-hour-and-50-minute lunar day.

The relative positions of the earth, moon, and sun also modify tidal height throughout the month. Twice a month, at the new and full moon cycles, when the earth, moon, and sun are all in alignment, the combined gravitational pull of the moon and sun causes higher-than-average tides called spring tides. These spring tides have nothing to do with the annual season of spring; rather, they spring up 20–30 percent higher than average high tides. Each spring tide lasts about four days. When the moon and sun are completely out of phase, during the first and last quarters of the moon, tidal ranges are 20–30 percent lower than average, and the resulting unusually moderate tides are called neap tides.

Tides are also influenced by how close the earth is to the moon. The moon does not rotate around the earth in a perfect

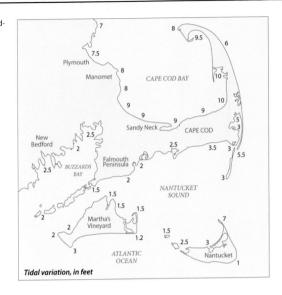

Tidal ranges vary greatly depending on location around the Cape and Islands region.

Tidal variation, in feet

circle: it rotates in a slightly oval path that puts the moon closer to the earth twice every 27.3 days. When the moon is closest to the earth in its oval orbit, it is said to be in perigee. Roughly twice a year the occurrence of spring tides in the new or full moon phases will coincide with the moon's closest approach to earth in its orbit, and we get extremely high tides, called perigean spring tides or sometimes king tides.

Winds can also affect tide cycles. In shallow waters a strong opposing wind can temporarily slow or even stop tidal flow. The nightmare scenario for weather affecting tides is when a hurricane or nor'easter arrives at the same time as a spring tide or—even worse—a perigean high tide. The combined high tides and storm surge can cause terrible coastal flooding. In April 1940 a nor'easter arrived during a perigean high tide and drove water 13 feet above the normal high tidemarks, flooding many New England and Long Island coastal towns and causing extensive damage.

The shape and depth of bodies of water and the surrounding landforms also affect the range of tidal movements. The tidal ranges of ports around the Cape and Islands vary greatly (see illustration, above). In Nantucket Sound the tidal flows are slowed by the many large and small islands surrounding the sound, and tidal ranges on the islands and the Cape's south shore are relatively small, typically one to three feet. In contrast, the tidal ranges on Cape Cod Bay are seven feet to ten feet in Wellfleet Harbor. Several factors account for these higher tidal ranges. No large islands exist to slow tidal movements in the more open bottom topography of Cape Cod Bay and the even larger Gulf of Maine. The bay and Gulf of Maine are also well exposed to the ocean beyond, where the tidal range is about six feet on the ocean side of Cape Cod. Further, the semienclosed shapes of the Gulf of Maine and Cape Cod Bay act as resonant tidal basins. In such enclosed bodies of water, the daily in-out movements of the tides set up a recurring wave of momentum or resonance (think of a swing moving back and forth), amplifying the tidal movements in both high- and low-tide cycles. The effect of tidal resonance is most pronounced in very enclosed spaces such as Provincetown and Wellfleet Harbors.

Marsh Elder (*Iva frutescens*), also known as High Tide Bush, is a good indicator of the Mean Spring High Water mark, the highest level of saltwater flooding in salt marshes, which happens twice each month. Marsh Elder is *very* common in salt marshes and on protected beaches, and so makes an easy landmark for the high tide line.

Zonation

All coastal marine environments are organized in vertical zones. Tide levels and the slope of the land as it meets the water determine the zones of marine life in salt marshes, on shoreline rocks, and on beaches, where the difference of a few inches of tide level or degrees of slope can completely alter the vegetation and animal life in a zone.

Coastal environments also have horizontal zones. The most common factors in horizontal zoning are slope and the distance to freshwater. Salinity and freshwater flow also influence which plants and animals can survive in a particular habitat.

Salt spray is another horizontal zoning factor. Only a few plant species can thrive in areas that regularly receive salt spray, so distinct zoning patterns form along coastline vegetation. These zones separate areas that receive a constant spray or dusting of salt from others that are more sheltered from wind-borne salt.

Groundsel Tree (*Baccharis halimifolia*) occurs on the edges of salt marshes and protected beaches, just above the high tide level, usually near or just behind the line of Marsh Elders.

Too much tidal soaking can drown some plants and animals or bring in too much salt. Too little exposure to tides starves many marsh creatures like fiddler crabs and mussels. Marshes and tidal flow provide important shelter to the young of many species of fish and marine invertebrates. The influence of tidal salt water is especially obvious in salt marshes. In the low marsh that is partly submerged twice a day, tall, salt-tolerant grasses like Saltwater Cordgrass predominate. Higher in the marsh, where plants are less exposed to salt water, shorter grasses like Saltmeadow Cordgrass form large salt meadows.

Knowledge of average high and low tidemarks is usually sufficient for a quick understanding of most coastal environments, but in salt marshes the monthly variation of spring and neap tides ultimately controls zonation. Luckily, you don't need exotic tide tables to see the zones; you just need to look at the pattern of plants to infer how high the highest

The recharge to and discharge from the Cape Cod aquifer

1– Precipitation as rain or snow
2– Evaporation from the ground
3– Evaporation from pond surfaces
4– Streams running to the ocean
5– Spring flow to the ocean from water table
6– Subsurface loss of groundwater
   to the ocean

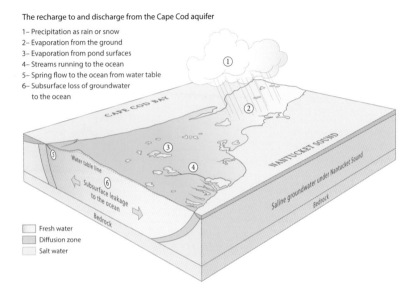

Fresh water
Diffusion zone
Salt water

high tides get (the Mean Spring High Water, or MSHW, level). In a typical Outer Lands salt marsh, the MSHW level will be marked by Marsh Elder (also called High Tide Bush) and Groundsel Trees, which will grow right to the edge of a salt marsh but cannot tolerate much direct contact with salt water. Spot those two bushes, and you'll know how high the highest normal tides get in that marsh.

Groundwater in the Outer Lands

Given the Outer Lands' porous, sandy soils, limited land area for drainage into groundwater aquifers, lack of rivers, and surrounding ocean waters, it seems remarkable that the region has much fresh groundwater at all. Without a significant reservoir of groundwater the Cape and Islands region would be almost uninhabitable. Rainfall and—to a much more limited extent—snowmelt are the only sources of freshwater in the Outer Lands. Luckily, two factors combine to provide freshwater in these salty surroundings.

Opposite:
The Cape's groundwater is not evenly distributed across the landscape, but is held in six major groundwater "lenses," including the very small Pilgrim Lens under the Provincelands dunes. Nantucket, Martha's Vineyard, Block Island, and New York's Long Island are all large enough to form groundwater reservoirs similar to those of Cape Cod.

On Cape Cod and the islands of the Outer Lands the fresh drinking water originates entirely from rain or snow. Most precipitation is rapidly absorbed by the sandy, porous soils, where it can be pumped to the surface for human use.

Freshwater is lighter than salt water, and in the ground a body of freshwater will float above a surrounding body of salt water, preventing a mixing that would make the resulting brackish water undrinkable. The sandy soils quickly absorb

Geologists refer to groundwater pools as lenses because of their shape: thick in the middle of land areas, tapering to thin edges near the shorelines. Martha's Vineyard, Nantucket, Block Island, and Long Island are all large enough to provide absorption and drainage aquifers.

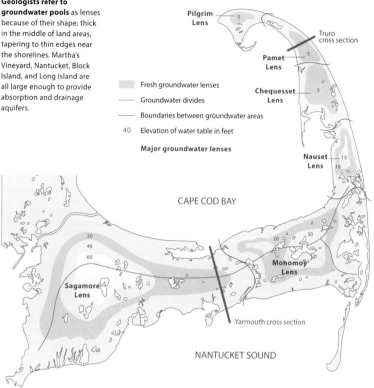

Pilgrim Lens

Truro cross section

Pamet Lens

Chequesset Lens

Nauset Lens

15

10

Fresh groundwater lenses

Groundwater divides

Boundaries between groundwater areas

40 Elevation of water table in feet

Major groundwater lenses

CAPE COD BAY

20
40
60

20 30

Monomoy Lens

Sagamore Lens

Yarmouth cross section

NANTUCKET SOUND

Yarmouth cross section

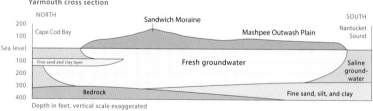

NORTH   SOUTH
200
100   Cape Cod Bay   Sandwich Moraine   Mashpee Outwash Plain   Nantucket Sound
Sea level
100   Fine sand and clay layer   Fresh groundwater   Saline ground-water
200
300   Fine sand, silt, and clay
400   Bedrock

Depth in feet, vertical scale exaggerated

Truro cross section

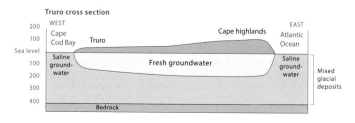

WEST   EAST
200   Cape Cod Bay   Cape highlands   Atlantic Ocean
100   Truro
Sea level
100   Saline ground-water   Fresh groundwater   Saline ground-water   Mixed glacial deposits
200
300
400   Bedrock

rainwater from the ground surface, but the porous, mixed glacial soils of the Outer Lands also hold significant quantities of water for long periods, acting both as a giant underground sponge and as a huge natural filter to clean impurities from the groundwater. Thus the ground under the Cape and Islands acts as a giant invisible freshwater reservoir—a much larger store of water than any of the visible lakes and ponds in the landscape. Geologists call these groundwater pools lenses owing to their shape: thick in the middle of land areas and tapering to thin edges near the shorelines. Martha's Vineyard, Nantucket, Block Island, and Long Island are all large enough to provide absorption and drainage aquifers, and the overall dynamics of the water cycle on the major islands is similar to that on Cape Cod.

Groundwater on Cape Cod is not evenly distributed; it is divided into six groundwater reservoirs, each one a finite resource that is only slowly replenished. Unfortunately, all of these groundwater sources are under stress due to the enormous demands of commercial development and new housing on the Cape. As of the 2014 federal census, the Cape had a permanent resident population of about 214,500, and that doubles to about 450,000 during the summer travel season. Of the roughly 263,000 acres of land on Cape Cod, about 44 percent had been developed by 2010. Every year more of the Cape's natural ground cover disappears under parking lots, driveways, and home lawns and septic systems that leak nitrogen (from lawn fertilizers), phosphorous (largely from home soaps and cleaning products), and other pollutants into the groundwater and surface ponds. Impervious surfaces like roads, building roofs, and parking lots also impede the absorption of rainwater and snowmelt into the groundwater supply.

The Upper Cape around Joint Base Cape Cod also faces a significant groundwater pollution problem resulting from many decades of poor handling of fuels and oil lubricants. A huge underground plume of contaminated groundwater continues to drift south from the military base, potentially threatening the safety of drinking water supplies in the areas around Mashpee and Falmouth.

Silver Spring Pond at the Massachusetts Audubon Society's Wellfleet Bay Wildlife Sanctuary. The narrow valley around the pond was formed as a meltwater channel in the Wellfleet outwash plain. A small dam created the pond, which is fed by spring seepage from the Nauset groundwater lens (see illustration, p. 69).

Mayflower II, Jim Curran

Environmental History

Shawme Lake, Sandwich. Sandwich was established in 1637 by settlers from Saugus, Massachusetts. Sandwich sits on rich lake-bed sediments from Glacial Lake Cape Cod Bay, which were ideal for farming.

The human effects on the Outer Lands began thousands of years in the past, when early Native Americans moved into the region perhaps 10,000–11,000 years ago. Our human life-times are short compared to the rate of geologic or ecological change, so we tend to think of our natural environments as relatively fixed. But whether we look at dates in the geologic history of the region, the effects of early humans, the begin-nings of permanent European settlement in 1620, or the Cape and Islands as they are now, each is just a snapshot in a continually evolving landscape. Today you can hardly escape the headlines about the earth's rapidly changing and warm-ing climate. Change itself is not new: our climate, average temperatures, and the sea level have always been changing. It's the accelerating *rate* of human-caused change that wor-ries climate scientists and anyone else who values our wild environments.

The environmental history of the Outer Lands can be sum-marized in three phases after permanent European settlement began in 1620, and each has left its mark on the Cape and Islands region we see today.

The era of resource exploitation (1620–1820), when early European settlers and later Americans extracted the environ-mental riches of the region through agriculture, lumbering, fishing, and coastal whaling.

For a concise political and cultural history of Cape Cod, see: *A Short History of Cape Cod,* by Robert J. Allison. Beverly, MA: Commonwealth Editions, 2010.

A bronze bas-relief plaque at the base of Provincetown's Pilgrim Monument commemorates the signing of the Mayflower Compact, the first governing document of Plymouth Colony. Below, a replica of the Pilgrims' ship, *Mayflower II*, sails on Cape Cod Bay.

Environmental history, a discipline that has emerged over the past three decades, looks at how human cultural and political history influences—and is influenced by—the natural environment. I highly recommend: *Cape Cod: An Environmental History of a Fragile Ecosystem,* by John T. Cumbler. Boston: University of Massachusetts Press, 2014. This is a comprehensive look at the human history of Cape Cod and how human activity and the natural environment have interacted over the past 400 years.

The decline of local resources (1820–1900), when the failure of local farms and a limited supply of local wood, combined with the rise of more efficient food production in the Midwest, forced the people of the Outer Lands to seek their livelihoods in fishing, large-scale whaling, and shipping, often in regions of the world far from New England.

The rise of the tourism economy (1900–present), when the seasonal tourism-based economy we know today originated, slowly before World War II and rapidly following 1945.

The Outer Lands before European settlements

Early European explorers and settlers in the Outer Lands found thick old-growth forests covering much of the region that were far older and more biologically diverse than any current forests on the Cape, Martha's Vineyard, or eastern Long Island. *Mourt's Relations,* an early account of the landing of the Pilgrims first at Provincetown and later at Plymouth, tells of an Outer Cape Cod that was "so goodly a land, wooded to the brinke of the sea." At the first Pilgrim anchorage, in Provincetown, "we came to an anchor in the bay, which is a good harbor and pleasant bay, circled round, except in the entrance which is about four miles over from land to land, compassed about to the very sea with oaks, pines, juniper, sassafras, and other sweet wood." The Pilgrims made no mention of large sand dunes in the Provincelands around what is now Provincetown. They described a land "covered with a tolerable growth of pine wood, shrubbery, and beech." However, a crew member of Bartholomew Gosnold's voyage of 1607 described the tip of Cape Cod as "only a headland of high hills of sand overgrown with shrubby pines, hurts [blueberries], and

Herring River, Wellfleet. Today's very young forested shorelines can only hint at what the Pilgrims found in their first explorations of Cape Cod in 1620. The Pilgrims found thick ancient hardwood forests, full of oaks, beeches, and White Pines, far taller and more diverse than the mostly Pitch Pine forests around the Outer Cape today.

such trash, but an excellent harbor for all weathers." Both accounts are probably accurate, as even today the Provincelands contain rich American Beech forests, sections of sand dunes largely barren of vegetation due to wind erosion, and sand plain forests dominated by Pitch Pine. What is more certain is that both the hardwood and sand plain forests were far more ancient and complex than anything we see in the Provincelands today. Echoes of this vanished forest appear in place-names: Today Wood End is a low sandspit of dunes and a lighthouse at the western end of the Provincelands, distant from any forest, but in 1620 it was literally where the woodlands ended.

The forests of precontact Cape Cod nevertheless contained the same basic mix of tree and shrub species we see in the landscape today, so we can make educated guesses about how those ancient forests looked based on both prehistoric pollen evidence and modern woodlands. On eastern Long Island lake-bed pollen studies show that dry-adapted coastal forests dominated by Pitch Pine, White Oak, Black Oak, Northern Red Oak, Bear Oak, and Scarlet Oak covered the sandy outwash plains there in the time just before European colonization. On higher ground along the regional moraines of the Outer Lands the forests were dominated by the oaks, plus scattered individuals of American Beech, American Chestnut, and Pignut Hickory.

The one major exception to the mostly vanished Outer Lands primeval forests is the Long Island Pine Barrens, where the dwarfed Pitch Pines were too small and thinly scattered to attract many woodcutters and the heath and shrub ground cover were not suitable for grazing livestock or conversion to farmland. The Pine Barrens' core preserve of 52,500 acres south of Riverhead, Long Island, is by far the largest intact wilderness area in the Outer Lands. See the "Coastal Forests" chapter (pg. 293) for more on this unique habitat.

Native American land use
After the glaciers retreated from the Outer Lands about 15,000 years ago, Paleo-Indian–Period hunter-gatherers likely moved into what was then a fairly harsh, tundralike landscape. The first radiocarbon-dated evidence of human activity in southern New England dates from about 10,000 years ago. At that time the sea level was 150–200 feet lower than it is today, and early people probably moved into the ice-free refugium areas on what is now the continental shelf south of the Cape and Islands and Long Island, but evidence of those early human settlements has long since been submerged by the Atlantic Ocean.*

*The long, complex history of the Native American peoples of southern New England is beyond the scope of this guide. For more information, I highly recommend these three books:

The Saltwater Frontier: *Indians and the Contest for the American Coast,* by Andrew Lipman. New Haven: Yale University Press, 2015.

Connecticut's Indigenous Peoples: *What Archaeology, History, and Oral Traditions Teach Us About Their Communities and Cultures,* by Lucianne Lavin. New Haven: Yale University Press, 2013.

Changes in the Land: *Indians, Colonists, and the Ecology of New England,* by William Cronon, revised edition. New York: Hill and Wang, 2003.

Overleaf:
Blackwater Pond at Beech Forest, Cape Cod National Seashore. If you have a hard time imagining the sandy tip of the Cape covered with rich forests, take a stroll through the Beech Forest area just outside of Provincetown. This lush, mixed hardwood and conifer forest with freshwater ponds is based entirely on dune sand.

During this time the climate in the Outer Lands region was rapidly warming, and the area was transitioning from tundra and spruce-fir boreal forests into the oak-maple-hickory eastern deciduous forests of today. When the first European explorers arrived in coastal New York and southern New England, the area had approximately 70,000 Native American residents, but population estimates are imprecise because so many Native Americans died of introduced European diseases before any practical census could be conducted. By 1675 the Native American population of New England had fallen to fewer than 12,000 individuals.

Most Native Americans in the Outer Lands region spoke closely related variants of the Algonquian language and had distinct territorial areas along the New England shoreline and eastern Long Island. Coastal tribes migrated between shoreline settlements in warmer weather and more sheltered inland locations during winter. The Native American tribe and place-names that survive were assigned by European settlers who phonetically translated the original Algonquian names. Those ancient names are some of the most familiar in the region: Nauset, Nantucket, Mashpee, Naushon, Chappaquiddick, Narragansett, Quonset, Montauk, and Shinnecock are but a few of the Native American tribe and place-names now famous in the region.

Native Americans in southern New England and the Outer Lands region are known to have used fire extensively to clear undergrowth in favorite game hunting areas and land for planting crops. The open woodlands, with patches of open grassland, were especially appealing to the White-Tailed

Native American groups of the Outer Lands and southern New England region.

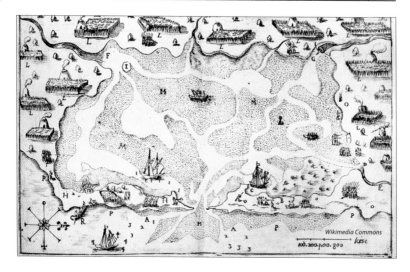

Wikimedia Commons

Deer, a major game animal for the Native Americans. Thus Native Americans were not passive actors in a pristine natural landscape—people had been influencing the southern New England landscape for 10,000 years before the Pilgrims arrived in 1620. However, because of their relatively small and scattered populations, the overall ecological impact of Native American settlements, agriculture, hunting, shellfish gathering, and fire use was environmentally benign compared to what came after Europeans arrived. Except near large villages and permanent settlements along the coast in places like the Boston peninsula, Eastham, the Chatham area, and around Narragansett Bay, and aside from commonly using fire to clear forest undergrowth, Native American land use made relatively few widespread or long-term changes in the natural environments of the Outer Lands.

As European fishermen and explorers began contact with the New World in the 1500s, they introduced Eurasian diseases that immediately began to devastate Native American populations along the Atlantic Coast of North America. Because of long isolation from Eurasian diseases, Native Americans had little genetic immunity to smallpox, plague, influenza, and other diseases common among the crews of early explorers and fishing vessels from Europe. Even before permanent European settlers such as the Pilgrims and Puritans arrived, the coastal populations of Native Americans had been reduced by epidemics of Eurasian diseases caught from the European crews of fishing fleets and early explorers. As a result, it is difficult to gauge the ecological impact of Native peoples: by

French navigator and cartographer Samuel de Champlain sailed into what is now Nauset Harbor in Eastham in 1605. Champlain noted about a dozen Native settlements and farm fields surrounding the harbor, with about 150 inhabitants in total.

the time European settlers began recording histories, a large portion of the Native population was already gone. Violent conflict between Native Americans and early European settlers also took a heavy toll. King Philip's War of 1675–76 was perhaps the deadliest war (per capita) in North American history, and in its aftermath the victorious English settlers exiled many New England Native Americans to slavery on Caribbean sugar plantations.

The era of resource exploitation, 1620–1820

In the decades after King Philip's War, Pilgrims from the Plymouth area and Puritan settlers from areas around and north of Boston began to settle Cape Cod, Nantucket, and Martha's Vineyard. It's no accident that such early Cape Cod settlements as Sandwich and Barnstable were along the coast of Cape Cod Bay, where deep, rich, and relatively stone-free sediments from Glacial Lake Cape Cod Bay (see illustration, pp. 26–27) made the soils ideal for European-style agriculture. The large salt marshes of the Sandwich coast and behind Barnstable's Sandy Neck were also useful for grazing cattle and sheep.

Early in the colonial period, the high plains of Eastham and Wellfleet were clear-cut for fuel and farmland. European-style farming practices of the 1700s quickly depleted the sandy, fast-draining soils, and by the mid-1800s these Outer Cape areas were virtual wastelands of treeless abandoned farms. In the twenty-first century these high plains have recovered somewhat to become heaths and young Pitch Pine forests.

European farming methods and land use were fundamentally different from the forms of small-scale agriculture practiced by Native peoples in the Outer Lands. Native Americans had no domestic animals besides dogs, and they grew mixed-species crops of corn, beans, and squash in relatively small areas of cleared forest and grassy fields. All of the ground preparation, planting, weeding, and harvesting was done by hand. Native peoples living along the coastline relied heavily on shellfish for protein in their diets, harvesting the abundant supplies of oysters, clams, mussels, and lobsters.

In contrast, European farming methods required much larger fields of cleared land, and draft animals such as horses and oxen were used to plow large fields that were planted with single-species crops of corn or wheat. In addition to draft animals, European farmers kept cattle, sheep, and pigs, which in the early colonial era were often free to roam forests and salt marshes in search of fodder. The Puritan settlement of Boston was growing rapidly in the late seventeenth century, and demand for meat and grains encouraged the growth of farming towns along the southern shore of Cape Cod Bay.

Problems with domestic animals and land clearance

The free-roaming domestic animals imported by European settlers caused almost immediate problems throughout the Outer Lands, in damage to Native American crops and wild landscapes. Whereas Europeans usually protected their farm fields with sturdy fences, Native American crops traditionally were not fenced. Settlers' pigs, sheep, goats, and cattle could quickly destroy Native crops, and this damage increased tensions between the Europeans and their indigenous neighbors. Free-roaming and semiferal pigs, goats, and sheep also stripped fragile grasslands, heaths, and dune areas, exposing bare soil and sand to erosion.

In the Provincelands north of Provincetown, the combination of clear-cutting dune forests for fuel and building materials and allowing unrestricted grazing of free-roaming domestic animals created the Cape's first well-documented environmental crisis when the newly bared sand dunes began to encroach on Provincetown. By the early 1700s so much of the natural vegetation of the Provincelands had been stripped away that huge sand dunes began burying buildings on the outer edge of town, and sand also began to fill in the harbor, threatening the town's shipping and fishing businesses. In 1714 the colonial Massachusetts government forbade further grazing or land clearance in the Provincelands, but townsfolk largely ignored the ordinance. By 1729, however, the people of Provincetown had become so alarmed over the mountains of sand descending on their town that a stronger ordinance was issued and they began planting Pitch Pines and American Beachgrass on the barren dunes and putting up fencing to hold sand and restrict grazing. Thus much of the Pitch Pine forest you see today in the Provincelands is a partially restored and very young echo of the Outer Cape's original dune forests, much thinner and less diverse than the dense dune forests the Pilgrims found in 1620.

Modern Pitch Pines in a losing battle against the movement of loose dune sand in the Provincelands. Three hundred years after the early European settlers stripped the Provincelands of their vegetation cover, the ecosystem is still recovering, and large areas of bare sand remain.

A Pitch Pine stand in today's Provincelands, near the Beech Forest site. Most of the Pitch Pines here are relatively young trees planted in the nineteenth and twentieth centuries to stabilize the sand dunes. The older and denser stands now approximate the appearance of the natural pine forests when the Pilgrims landed in 1620.

Poor land use also plagued the inhabitants of Nantucket and Martha's Vineyard. Although Nantucket's early "rich forests of oak and pine" were probably never as thick or tall as mainland forests due to constant salt spray, what woodlands the island did have were quickly cut. Intensive farming quickly exhausted Nantucket's thin, sandy soil. In the 1720s an acre of Nantucket farmland could yield as much as 250 bushels of corn, yet by the time of the Revolution, that acre was lucky to yield 20 bushels. Excessive livestock grazing doomed much of the grasslands and heaths of Nantucket as well as of central and southern Martha's Vineyard. A visitor to Martha's Vineyard noted in 1763 that about 20,000 sheep grazed the central fields, and in 1888 Harvard geologist Nathaniel Shaler observed that about 33,000 acres on Martha's Vineyard had transitioned from verdant forest to fertile farmland and then to useless near-barren fields over a century of poor farming practice and excessive grazing.

Use of forests

In the colonial and early American eras (1620–1820) local forests were extensively cut to create farmland, provide building materials and heating fuel, and fuel such early local industries as brickyards, glassworks, and ironworks. The large hardwood forests of the Upper Cape remained fairly intact well into the mid-1800s, but the Outer Cape, Martha's Vineyard, and Nantucket faced wood shortages as early as the 1670s.

From 1620 until the early 1800s most regional homes were heated with an inefficient open hearth that required as many as 30 cords of hardwood for one New England winter. To be self-sustaining a farm might need a woodlot of 30–40 acres in addition to farm fields or grazing land. In the early 1800s most homes and businesses converted their heating to far more efficient wood stoves, which could heat an average home with 10–15 cords of wood per winter. The drop in wood consumption made large woodlots unnecessary, and most farmers clear-cut their woodlots to add land for crops or grazing.

The primeval forests of the Outer Lands disappeared by 1900. Most Cape and Islands woodlands have been clear-cut at least twice since 1620, and many have been cut three times. In addition to the loss of forest species diversity, clear-cutting exposed the topsoil to erosion and heavy grazing by cattle and sheep. By the mid-1800s many former woodlands had become open grassland and heath, with poor, thin soils often exhausted of nutrients through erosion and constant agricultural use.

In the mid-1700s early travel writers such as Timothy Dwight and Edward Augustus Kendall remarked on the rich woodlands of the Outer Cape. Yet by the time Henry Thoreau made his four trips through the Cape (1849–55), chronicled in *Cape Cod*, he described much of the Outer Cape from Orleans north to Truro as a "desert." Looking from Truro north to the Provincelands, "there was not a tree as far as we could see." Ironically, Thoreau admired the treeless view for the distant vistas it offered: "The almost universal bareness and smoothness of the landscape were as agreeable as novel, making it so much the more like the deck of a vessel. We saw vessels sailing south into the Bay, on the one hand, and north along the Atlantic shore, on the other, all with an aft wind."

Anadromous fish and dams

Cape Cod has no rivers of any great size or length, and many of the region's so-called rivers are long tidal inlets or estuaries rather than substantial flowing rivers. Rivers are absent because the landmasses of the Cape and Islands don't provide a watershed large enough to supply a river reliably and the sandy ground absorbs water quickly. The only sizable exception is eastern Long Island's Peconic River, and even the Peconic is sluggish and tidal for much of its length.

However, even the modest rivers and streams of the Outer Lands are important breeding grounds for anadromous fish such as herring, alewives, and shad. Anadromous fish are born in freshwater, swim to the ocean for their adult lives, and then return to freshwater rivers to breed. Most of the Atlantic

Anadromous fish of Cape Cod. The small mill dams in the Outer Lands region were valuable for processing local grain and to support early manufacturing, but the dams were often lethal for fish species that spawn in rivers, live their adult lives in the ocean, and then return to their ancestral rivers to breed. In recent decades New England towns have been removing old dams in an effort to revive anadromous fish migration into our freshwater rivers and streams.

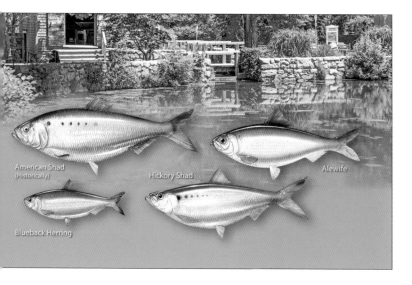

American Shad
(Historically)

Hickory Shad

Alewife

Blueback Herring

The historic Dexter Grist Mill, in Sandwich, built in 1654, is one of the oldest remaining water mills in the United States. The mill operated through the late 1800s. The mill site was created by damming the upper Shawme River, now the Shawme Lakes.

Coast's major sport and food fish are anadromous: American Shad, Hickory Shad, Alewife, and Blueback Herring are all anadromous, as are such larger game fish as Atlantic Salmon and Striped Bass. Outer Lands streams are also critical habitat for the American Eel, whose complex, multistage life cycle brings young, transparent glass eels into Cape and Islands freshwater streams in the spring and summer.

In the colonial era, water mills powered light industries such as grain and sawmills. The midsized streams of the Outer Lands were easily dammed to build mills. At first the mills observed the English common law obligation to provide bypass chutes or other accommodations to anadromous fish, but as time went on, mill owners ignored the law and many substantial streams on the Cape were fully dammed, in most cases dooming the spring runs of such anadromous species as herring and shad, which require fish ladders or alternate spillways around the mill wheels to bypass the dams. By the mid-1800s dammed rivers had decimated the once-abundant herring and shad breeding populations and blocked the upstream runs of American Eels.

Some Cape Cod streams still have substantial spring herring runs, usually of Alewives but also of Blueback Herring in small numbers. Stony Brook in West Brewster is known for its herring runs around the six or so weeks after the spring equinox; the stream is fed by a chain of small lakes that act as spawning areas. The main Herring Run Site (a public park) is on Stony Brook Road in Brewster at approximately 830 Stony Brook Road. Look for the historic grist mill and the roadside parking lot. The herring run strongly on days when the air temperature is above 50 degrees Fahrenheit. Watch for gulls: if the gulls are swarming, the herring are running.

Early commercial fishing

When Bartholomew Gosnold christened "Cape Cod" during his exploration voyage of 1602, the region had already had nearly a century of routine visits from Europeans. A few were early explorers, but most Europeans visiting the New World were Basque, French, and Portuguese fishermen coming for the vast schools of Atlantic Cod. The early Pilgrim and Puritan settlers were primarily farmers, but the Outer Lands region and its fine natural harbors quickly became famous for fishing. By the late 1700s more than 665 commercial fishing vessels left Cape harbors for regional fishing banks and even the Grand Banks off Newfoundland. Cape fishermen were initially wary of Georges Bank, which was notorious for its treacherous currents, high waves, and exposure to sudden storms and thick fogs.

Fish of all kinds were abundant, but in the days before refrigeration and ice makers, salting and drying was the only means to preserve the catch. In the early days salt had to be imported from Europe and was therefore expensive. During the Revolutionary War the British closed off the supply of salt, effectively shutting down large-scale regional fishing in New England. In 1776 a Dennis inventor, John Sears, built the Cape's first salt evaporation shed, where shallow pools of ocean water were left to evaporate, yielding good-quality sea salt in quantities large enough to supply local fishing fleets. Soon the shores of Cape Cod were dense with saltworks and flat wooden fish flakes, on which salted cod fillets were dried in the sun before shipping. By the 1830s the Cape was producing 669,064 bushels of salt annually. One observer noted that for 20 miles southeast of Provincetown the shoreline "seemed to be built of salt vats."

After the Wars of Independence and 1812 ended, the local waters were safer and the newly available and cheaper salt made it possible to process large quantities of fish, and commercial fishing in the Outer Lands vastly expanded. Provincetown's large natural harbor and location 24 miles off the mainland made it the largest fishing port in the region south of Gloucester. In addition to a seemingly inexhaustible supply of Atlantic Cod, fishermen caught vast numbers of Atlantic Halibut, Atlantic Mackerel, and such groundfish species as Pollock, Haddock, and Silver and White Hake.

A fish flake with drying salted cod fillets. The shorelines of fishing towns such as Provincetown were dense with saltworks and fish drying operations until the early twentieth century, when commercial refrigeration replaced salting and drying as a preservative.

Wikimedia Commons

The Essex-built
banks fishing
schooner "Columbia"

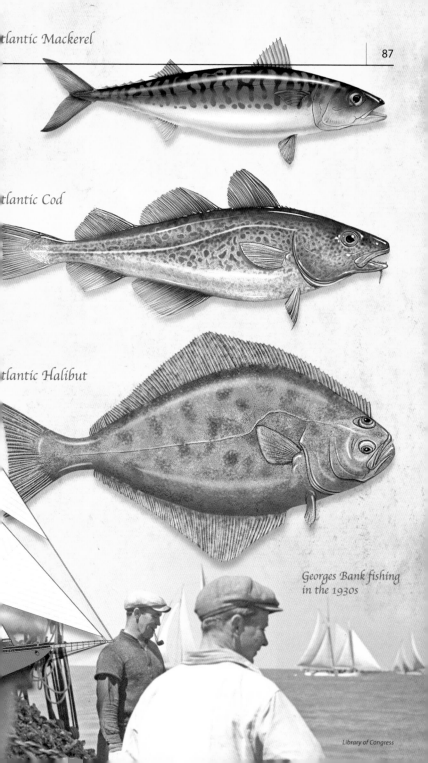

Georges Bank fishing in the 1930s

An innovation in fishing vessels made more ambitious voyages possible. Before the War of 1812 most New England fisherman were using sturdy but slow chebacco boats, well suited for near-shore fishing but awkward for open ocean use. Harassment by the Royal Navy and privateers also spurred interest in larger, faster hulls and rigging designs that could reach the offshore banks quickly and—most crucially—return to port quickly before the catch spoiled. By the 1820s Gloucester and Essex boatyards were building the first fishing schooners, twin-masted vessels with long, sleek hulls and huge sail areas that could reach the Grand Banks or Georges Bank quickly and cope with the rough seas far offshore. Other Outer Lands fishing ports quickly followed Gloucester's lead, and a century of fast, elegant fishing schooners began.

The fishing schooners were famously beautiful but also deadly. With their large spread of canvas and low-slung hulls built for speed, schooners were often lost in the stormy waters of the North Atlantic, especially when they were too heavily laden with fish. A single classic nor'easter, the gale of October 3, 1841, wrecked at least 190 vessels, most of them fishing schooners. On the Georges Bank at least 81 Cape Cod fishermen were drowned, most Cape Cod–based saltworks were destroyed, and the regional fishing industry went into a multiyear slump. A memorial in the churchyard of the Congregational church in Truro commemorates the loss of 57 Truro men and boys in the 1841 gale. However dangerous, the era of fishing schooners lasted for a century into the early 1920s, when motor vessels replaced schooners.

The decline of local resources, 1820–1900

The farms, mills, and light industry of Cape Cod and the islands continued through the mid-1800s, but the exhaustion of local farmlands coupled with the dwindling supply of wood made local agriculture and animal husbandry increasingly difficult. As much richer farmland in the Midwest became available, and new transportation systems such as canal networks and early railroads were developed, many Cape and Islands farmers simply left for greener pastures. By the early to mid-1800s sections of Outer Cape towns such as Orleans, Eastham, and the highlands of Wellfleet were desertlike expanses of abandoned and treeless farmland, then beginning to turn into the grasslands and heaths we see today in the Marconi Station and Highland Light areas of Cape Cod National Seashore, but much more open, with only small, shrublike Pitch Pines and Bear Oaks for trees.

As the riches of the land faded, the people of the Outer Lands increasingly turned to the sea for their livelihood. As expert seamen with almost two centuries of local maritime tradition behind them, men and boys from the Outer Lands provided officers and crew for the booming American shipping trade with Europe and the Far East. Commercial fishing continued as a mainstay of the Outer Cape Cod economy, but after the privations and dangers of the War of 1812, offshore whaling became the key regional industry between the War of 1812 and the beginning of the Civil War in 1861.

Whaling in the Outer Lands

Since the early days of the Pilgrims the people of the Outer Lands had valued whales for their baleen and rendered fat for candles and lighting. Through normal mortality in the large populations of whales in the waters surrounding the Outer Lands, dead or dying drift whales frequently washed ashore, providing a bonanza to local inhabitants, both European settlers and Native peoples. Oddly, although meat was precious in early colonial America, the European settlers generally refused to eat whale meat, and most meat on drift whales was left to rot on the beach. Pods of smaller whales such as Long-Finned Pilot Whales, Grampus, and dolphins also stranded regularly, particularly in the shallows of the eastern shores of Cape Cod Bay from Brewster north to Truro. All the peoples of the region were thus well aware of the value of whales, but through most of the 1600s only drift whales were used.

Whale hunting began gradually in the late 1600s and early 1700s, from processing drift whales on the beach to launching whaleboats from the beach to actively hunt local inshore species such as the Northern Right Whale and Humpback Whale. Critically for early whalers, Humpbacks and Right Whales were so fatty that they would float when dead. Fin Whales were also abundant in coastal waters, but Fin Whales are comparatively sleek and fast, and they sink when killed. Whales were harpooned from boats, and the dead whales were towed to shore for processing on the beach. Large tryworks rendered oil from blubber flensed from the whale carcass, and the rendered oil was sealed in barrels for transport.

In 1690 Nantucket Islanders invited Ichabod Paddock of Yarmouth, Cape Cod, to "instruct them in the best manner of killing whales and extracting their oil." By that time the native resources of Nantucket were depleted and islanders needed a means to support themselves. The forests had been cut down, crop yields from farming the thin, sandy soils were decreasing, and the 10,000 sheep roaming the island had reduced the natural heath and grassland environments to useless scrubland. In 1700 more than half of Nantucket's population was Native American. Native peoples had long experience in

Library of Congress

The New Bedford whaleship *Wanderer* of 1878, one of the last regional whalers. The *Wanderer* was lost in a gale in 1924, when she foundered and wrecked on Cuttyhunk Island, the outermost of the Elizabeth Islands.

Outer Lands Whaling

Humpback Whale

Sperm Whale

Right Whale

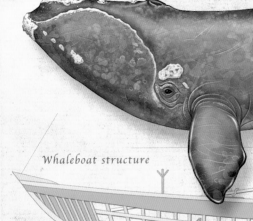

Whaleboat structure

Charles W. Morgan
Mystic, Connecticut

Replica of a Charles Beetle whaleboat,
Mystic Seaport

processing drift whales, but there is little firm evidence that they engaged in active offshore whale hunting before this time. However, the "Nantucket Indians" quickly became famously skilled hunters of whales, and as whaling transitioned from shore to ship, more than half the crews were Native Americans, mostly from Nantucket.

By the 1740s shore-based whalers from Nantucket, Provincetown, and Wellfleet and Long Islanders from ports like Sag Harbor had depleted local waters of whales, and they began outfitting larger ships for offshore whaling. Shore-based whaleboats blown far out to sea in gales had returned with news of Sperm Whales far offshore, in the New England canyons region at the edge of the continental shelf, where 400-foot-deep waters suddenly plunge to depths of 6,000 feet or more. Sperm Whales are deep divers that seldom enter the relatively shallow waters of the continental shelf, but the Nantucket whalers knew that these whales produced the best-quality oil and candles, so they had a powerful incentive to design whaleships capable of longer voyages. By 1748 Nantucket was home port to 60 substantial whaleships hunting Sperm Whales in Northwestern Atlantic waters from Newfoundland south to Bermuda and as far east as the Azores. On Cape Cod the only ports deep enough for larger whaleships were Provincetown and Wellfleet, which also outfitted ships for offshore whaling, but not in the numbers that Nantucketers did.

Nantucket prospered as North America's premier whaling port until the Revolution (1775–83). Hit hard by British blockades and whaleship seizures, Nantucketers struggled during the war, as did all American whalers. The War of 1812 (1812–15) was a further blow, when many American whaler crews were captured and impressed into the Royal Navy. After the wars, however, all American whaling ports enjoyed a remarkable run of prosperity. The American harvest of whale oil soared from about 30,000 tons in 1815 to almost 240,000 tons in the best whaling year, 1848.

Whaling and the Civil War

The California Gold Rush of 1849 had a negative effect on whaling. Many whaleships were converted to passenger ships for the passage around Cape Horn to the West Coast, and many former whalers jumped ship in San Francisco to become gold miners, leaving the whaling fleet undermanned. The huge increase in the American whaling fleet was paralleled by the whaling fleets of England and other European countries, putting enormous pressure on Sperm Whale populations throughout the globe. But it was the Civil War

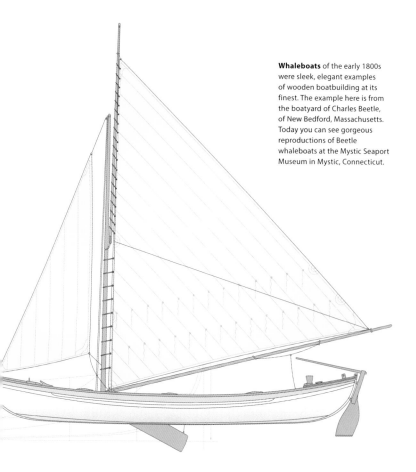

Whaleboats of the early 1800s were sleek, elegant examples of wooden boatbuilding at its finest. The example here is from the boatyard of Charles Beetle, of New Bedford, Massachusetts. Today you can see gorgeous reproductions of Beetle whaleboats at the Mystic Seaport Museum in Mystic, Connecticut.

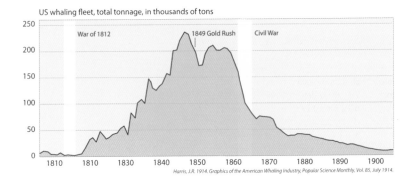

US whaling fleet, total tonnage, in thousands of tons

that signaled the sharp decline of the American whaling fleet. The outbreak of hostilities drew crews away to the fighting, and ship owners found it increasingly hard to get insurance for their voyages, as Confederate privateers freely roamed Atlantic waters in search of northern ships. In late 1861, the US Navy assembled what became known as the Stone Fleet, purchasing various old ships, most of them whaleships, filling them with granite boulders, and sinking them at the entrance to Charleston Harbor in South Carolina in a vain attempt to blockade the harbor. This sinking ultimately failed, as did a second Stone Fleet in 1862. During the war the Confederate fleet deliberately targeted whaleships, and by war's end, the American whaling fleet had lost 80 whaleships.

The decline of whaling

The destruction wrought by the Civil War nearly ruined the American whaling industry. The fleet might have rebounded with new ships after the war but for two factors that accelerated the decline in whaling's fortunes. In 1859 Edwin Drake struck oil in western Pennsylvania, and when whale oil grew scarce during the Civil War, petroleum-based kerosene quickly replaced Sperm Whale oil for home lighting. Although consumers often preferred whale candles over kerosene lamps, a second and ultimate problem became evident to even the hardest-driving whalers: the whales were almost gone, and multiyear voyages to distant Pacific, Arctic, and Antarctic waters were becoming too expensive to be profitable. Advances in industry brought many new, good-paying jobs that did not require the dangers and long voyages of whaling. Men left the whaling industry and moved their families away from the Outer Lands, and the populations of the Cape and Islands began a long decline that did not significantly reverse until after World War II.

Changes and trouble with commercial fishing

The decline in whaling coincided with a reduction in Atlantic Cod populations in the local waters of the Outer Lands. As inshore stocks of cod disappeared, the new, larger, more seaworthy fishing schooners made longer trips out to the Grand Banks and Georges Bank in search of cod, which were still abundant offshore—at least some of the time. Cod stocks could vary widely from year to year, and after the Civil War, Outer Lands fishing began to target schools of Atlantic Mackerel. For a few decades the mackerel supply held, but predictably the unrestricted fishing pressure caused a fall in mackerel stocks, and by the 1880s the mackerel catch was in steep decline. Cape Cod fishing ports and fleets began to shrink steadily. In 1850 Wellfleet supported a fleet of 100 ships in whaling, fishing, and coastal trade, but by 1890 only 20 remained, and the town's population shrank accordingly.

In the early 1920s steam-powered and later diesel-powered beam trawlers replaced the older fishing schooners, and the increasingly larger catches enabled by power vessels with better equipment began to imperil the fish stocks of the region in a way that the schooners and handline catches from dories never could in their day. The trawlers were so efficient at catching fish that even the fishermen began to worry that too many fish were being taken from New England waters. Cod were becoming scarcer, and the major trawler fleet switched to the still-plentiful Haddock (a cod relative), but after about 20 years of unrestricted fishing, Haddock also became scarce. Nevertheless, Cape fishermen persisted, and Provincetown continued as a major New England fishing port, expanding in population as other Cape and Islands fishing ports faded and

A drag trawler in Provincetown Harbor, 1942. Powered fishing vessels using dragnets were far more efficient than the sail-powered schooners they replaced, but their very effectiveness endangered regional fish stocks.

Library of Congress

Mackerel fishing on Georges Bank in the 1930s.

gradually lost their fishing fleets. Chatham had been a major fishing center, but continual problems with the shifting sand bars around Chatham Inlet made it difficult to run larger off-shore fishing vessels from the port, and the larger craft moved to Provincetown's large, deep harbor. Wellfleet had also been a major whaling and fishing center, but poor land management caused much of the harbor to fill with silt and sand, and it became too shallow for larger vessels.

Portuguese families from the Azores who had immigrated to work in the whaling industry shifted to commercial fishing as whaling faded. By the 1890s people of Portuguese Azorean descent made up a quarter of Provincetown's population and the majority of commercial fishermen. As local stocks of fish dwindled, New England fishermen made increasingly long voyages to the Grand Banks and Georges Bank, braving the stormy and dangerous North Atlantic to maintain sufficient catches to stay in business.

Eastern Long Island

English from Connecticut established several small communities on eastern Long Island by the mid-1600s, but they did not settle in western Long Island until after the Dutch surrendered New Amsterdam to the English in 1674. The English settlers on eastern Long Island considered themselves part of the Connecticut Colony, but the Duke of York, who resented Connecticut for having harbored the regicide judges who sentenced his father, King James I, to death, forced Long Island to join New York in 1676, ending Connecticut's claim to the island. During the Revolutionary War and well into the mid-1800s, Long Island remained largely rural and sparsely settled. Settlement gradually moved eastward from New York City, driven primarily by small farms that fed the growing city.

Because of the relatively late development of eastern Long Island, the region was spared many of the early environmental mistakes in agriculture and land use made on Cape Cod, Martha's Vineyard, and Nantucket. In the twentieth century, especially just before and after World War II, the north and south forks of Long Island saw a great deal of land clearance for agriculture and suburbs, but large natural areas of eastern Long Island such as the Pine Barrens are surprisingly intact.

The Long Island Rail Road, which reached Greenport on the north fork of the island in 1844, was planned as a faster, safer alternative to Long Island Sound steamship lines for passenger traffic and light freight shipping to Boston. Boston-bound New York passengers rode the Long Island Rail Road east to Greenport, where they took a steamship ferry across to

Stonington, Connecticut, and then boarded a train to Boston. As a financial venture the early railroad nearly failed, but the new access route across the island spurred development in the middle and eastern reaches of Long Island, and ultimately those new farming towns of the mid- to late 1800s proved to be the rail line's savior. During the mid-nineteenth century the railroad opened more than 50 stations in Nassau County and another 40 in Suffolk County, creating the foundation for the later growth of suburbs, farming, and industry in eastern Long Island.

The rise of the tourism economy, 1900–present

By the late 1800s the Cape and Islands terrestrial and marine environments were in poor condition following more than two centuries of exploitation and abuse. By 1850 less than 20 percent of Cape Cod was wooded. The Outer Cape was particularly barren and treeless, but even Mid-Cape towns that had been lushly forested in the 1700s, such as Dennis and Yarmouth, were now, in Thoreau's 1850s observations, "for the most part bare," "with hardly a tree in sight." Travelers' guidebooks to New England in the late 1800s described the Cape as a place of "widespread dissoluteness" and a "wilderness of sand." In 1889 the Cape's Old Colony Railroad called the Cape landscape "treeless, almost verdure-less, a barren plain, windswept and bleak." Ironically, the abandonment of many Cape and Islands farms and homesteads set the stage for a new era in the Outer Lands, where former farm fields began either to reforest or, in more exposed sandy areas subject to continual salt spray, to become heath and coastal grasslands, as seen today in the Marconi Station and Highland Light areas of the Outer Cape. North of Provincetown the planted areas of Pitch Pine began to stabilize the sand dunes, and the Beech Forest area evolved from small, isolated pockets of woods around Blackwater Pond to the fully developed, albeit relatively young, American Beech forest of today (see illustration, pp. 76–77). Many barren Outer Cape and Mid-Cape areas that had once been covered with mature forests began to reforest.

As the Outer Cape has recovered and much of the land has returned to some form of natural environment, the National Parks Service now has an ironic dilemma in its mission to conserve and restore the natural habitats of Cape Cod National Seashore. The barren and abandoned farm fields along the eastern edge of the Outer Cape between Eastham and Truro have largely converted to heath or heath mixed with scattered Pitch Pines. Heaths were a small component of the Cape geography at the time of the Pilgrims' arrival; historical records show that the Outer Cape was mostly coastal forest. But the heaths of today actually make the Outer Cape more

The Bourne and Sagamore Bridges (both opened in 1935) helped end the relative isolation of the Cape. Here the Bourne Bridge spans the Cape Cod Canal.

After the devastating deforestation of the Mid- and Outer Cape in the 1800s, most areas began to reforest, and today the Cape is about 80 percent forested.

ecologically diverse, and the Marconi heaths and coastal grasslands such as the areas around Fort Hill are now critical habitat for such uncommon species as Vesper and Grasshopper Sparrows and the Eastern Meadowlark. In the Anthropocene Age we live in today, where humans are the primary drivers of ecological change, what is and is not a natural environment has become a complex judgment call that must recognize both the human history and nonhuman influences that created the Outer Lands we know today.

Early tourism

The opening of the Cape Cod Railroad in 1848 and its eventual extension to Provincetown in 1873 and Chatham in 1887 were critical first steps in rebuilding Cape Cod as a vacation destination. The railroad was built primarily to haul Cape Cod fish and cranberries to the mainland, but along with produce and marketplace goods brought onto the Cape by the railroad, a small but steady stream of visitors began arriving. Before the railroad virtually all distant travel was by coastal packet boats and coastal steamers, which could be dangerous in winter and stormy weather. The railroad made journeys to the Cape safer and more convenient, and by the late 1800s people began to discover that the sleepy "wilderness of sand" was a charming place to spend a summer week away from crowded Boston or New York. By the early 1900s regular steamship service between Boston and Provincetown also began to transform the rough fishing village at the north end of Cape Cod into the summer tourist mecca we know today.

Despite its environmental and economic challenges at the end of the nineteenth century, the Cape had some advantages as a travel destination. The outmigration of Cape families after the collapse of whaling and the depression in local fishing left many abandoned homesteads and farms to return to grassland and forest, a turn away from the barren, depleted land of the mid-1800s. With many houses unoccupied, it was inexpensive to rent cottages or even the grand homes of former whaling officers. Since the mid-1800s Methodists had been gathering each summer for camp meetings, religious festivals and services that drew thousands of families to visit the Cape. At first the camp meetings were housed in tent towns erected in the vacant fields of such Outer Cape towns as Eastham, but as the tradition matured, families began to buy or build small yet more permanent summer homes and to extend their stays away from the crowded northeastern cities where most participants lived.

As Cape and Islands forests, heaths, and wetlands began to recover, wealthy sportsmen discovered the Cape. The first

Opposite:

The Boston-Provincetown steamship *Romance* in Provincetown Harbor, 1938. Regular steamship service from Boston made Provincetown a popular day-trip or weekend getaway from the summer city.

By the 1930s the influx of summer visitors began slowly to reverse the population slide that bottomed out in 1920, and as the nascent tourist trade grew, it could support more permanent residents on the Cape and Islands.

Paul Lemke

Woods Hole Harbor. Cape Cod has world-class bioscience and oceanographic institutions: Marine Biological Laboratory and the Woods Hole Oceanographic Institute.

hunting and fishing camps were rough and rural, but most gradually evolved into summer homes. The Upper Cape, particularly the south shore along Nantucket Sound, was convenient to the Boston and Providence, Rhode Island, areas, and began to attract wealthy families who summered on the cool Cape shores to avoid the sweltering cities in the days before air conditioning. In 1887 the US Department of Commerce's Bureau of Fisheries decided to form a research institution in Woods Hole, which already had several small scientific laboratories nearby. In 1888, 17 scientists and students formed the inaugural summer season of the Marine Biological Laboratory (MBL) in Woods Hole. The MBL grew rapidly in the next decade, and the annual influx of distinguished scientists, government researchers, and students helped transform the perception of the Cape as something other than a sandy backwater of fishing towns. Today the MBL and the associated Woods Hole Oceanographic Institute are world-class international centers for biomedical and oceanographic research.

Post–World War II and modern tourism

In 1914, Henry Ford revolutionized the young automobile industry by introducing the Model T Ford, the first car that middle-class Americans could afford to own. By 1925 the Ford Motor Company was producing 9,000 Model T cars a day, and the increase in automobile ownership in the 1920s led to the building of improved roads.

The first parkways in southern New England and Long Island date from the late 1920s and early 1930s, and along with better roads came development. The new roads created a demand for faster and more powerful cars. New transportation options allowed the rapid growth of affordable suburban housing in the 1950s and 1960s, particularly as Massachusetts Route 3 was expanded into a limited-access parkway heading from Boston to the Cape, and the Grand Central Parkway allowed easier access to central and eastern Long Island. In the early 1950s the Eisenhower-era Interstate Highway System Committee (dominated by Detroit carmakers) recommended that America's transportation needs be met almost entirely by cars and trucks. Construction of the Interstate Highway System began with the passage of the Federal Aid Highway Act of 1956.

The rapid growth of tourism after World War II immediately put pressure on Cape Cod's antiquated road system. By the late 1940s Massachusetts Route 6 (now Route 6A) and Route 28 were overwhelmed by summer traffic. In 1949 work began on a new, larger Mid-Cape Highway, the modern Route 6. In 1954 the new highway was expanded to four lanes from the

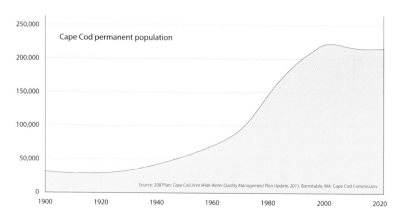

Cape Cod permanent population

Source: 208 Plan: *Cape Cod Area Wide Water Quality Management Plan Update, 2015.* Barnstable, MA: Cape Cod Commission.

Cape Cod Canal to Barnstable, and Route 28 was expanded to four lanes from the Bourne Bridge to Falmouth. On Long Island work began on the Long Island Expressway (Interstate 495) within New York City in 1940, and the highway was gradually extended east-west across the midsection of Long Island, finally reaching Riverhead in 1963. With the new roads in place, the Outer Lands were poised for the explosion of suburbs and strip malls and a steep rise in regional tourism and seasonal homeowners.

Troubles in vacationland

Today Cape Cod and the other areas of the Outer Lands are recognized as one of the premier vacation regions in the United States. Nearly 6 million people visit Cape Cod each year, with about 65 percent of those visits between June and September, and the resort and travel industry generates about 40 percent of the Cape and Islands region's employment. Hotel occupancy rates were 78 percent for July 2016 and 81 percent for August 2016, which is about average in the area for the past seven years, and strong compared to other vacation destinations. There were 4,532,768 visits to Cape Cod National Seashore in 2015. Although the Outer Lands may be in the best environmental shape they have been in since 1620, the explosion in building and tourism has led to problems over the years.

Paving paradise

In the decade between 2001 and 2011, Cape Cod lost more than 2,300 acres of woodland (about 2.5 percent of its total forest cover), 70 percent of which were replaced by home and business developments with some degree of impermeable ground cover. Aside from direct destruction and replacement

The reputation of the Cape and Islands as a vacation paradise cannot be taken for granted. The steady erosion of water and environmental standards will eventually cripple the Cape's reputation, as it has in the past.

of natural habitats with buildings, parking lots, lawns, and roads, one of the primary environmental concerns with the expansion of the built environment is impermeable surfaces, also referred to as impervious surfaces. Roads, sidewalks, driveways, parking lots, and large building rooftops are all impermeable surfaces, meaning that rainwater and snowmelt quickly runs off these artificial surfaces instead of soaking into the ground. In natural environments rain and snowmelt are absorbed into the ground, where contaminants can often be filtered and broken down by natural processes. Runoff into sewers and storm drains brings unfiltered pollutants directly into our municipal wastewater processing systems, and into our lakes, streams, and marine coastal environments. Runoff from towns and cities contains high levels of nitrogen and phosphorous from fertilizers, sewage, and household soaps, as well as a wide variety of general pollutants from house-hold and commercial chemicals, and automobile fluids and wastes. Many towns and businesses are working to help limit stormwater runoff through the use of rain gardens and other filtering and shunting systems, where runoff that might once have gone directly into sewers is instead shunted into rain gardens designed and planted specifically to filter and drain runoff water into the local groundwater.

Nitrogen and hypoxia

Nitrogen is a naturally occurring element and, along with car-bon, hydrogen, and oxygen, is a fundamental building block of life. In natural environments nitrogen is in high demand and is usually in limited supply and carefully conserved, particularly in plant physiology. This growth-limiting role of scarce nitrogen is especially important in marine and estuary ecosystems. When a large artificial supply of nitrogen—such as sewage or lawn fertilizer runoff—is introduced into an aquatic environment, the nitrogen accelerates the growth of simple, fast-reproducing algae, cyanobacteria, and diatoms, collectively called phytoplankton. Like all green plants, phytoplankton release oxygen as they photosynthesize, but at night or in low-light conditions phytoplankton use more oxygen than they release, and in their overabundance the phytoplankton can rapidly deplete the dissolved oxygen in polluted waters.

This condition of low dissolved oxygen is called hypoxia, and hypoxia usually follows large phytoplankton blooms. As the algae die from lack of oxygen, water conditions worsen because as the dead algae decompose, their tissues absorb what little oxygen is left in the water. Hypoxia is stressful for all aquatic organisms, and if it lasts too long or dissolved oxygen levels fall too low, hypoxia is lethal to both the phy-

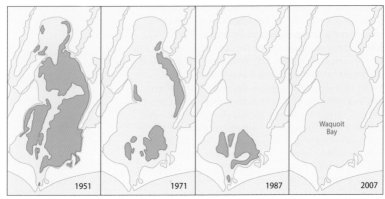

1951   1971   1987   Waquoit Bay   2007

After Costa, et al., 1992; Costello and Kenworthy, 2009.

The disappearance of Eelgrass beds in Waquoit Bay is due largely to excess nitrogen pollution of the bay's waters, primarily from septic systems and runoff from domestic lawns. Without the Eelgrass communities what was a highly productive estuary is no longer a major nursery area for local sport fish and shellfish.

toplankton and aquatic animals. Such overfertilized aquatic environments are said to be eutrophic—fertilized to the point where the system continually cycles through boom-and-bust sequences of rapid algae growth, mass algae death, and lethal hypoxia. Over the long term eutrophic systems become aquatic dead zones, where only the simplest phytoplankton can survive.

Many of the freshwater ponds on Cape Cod and the Islands show signs of eutrophication, as do small bays and inlets along the coasts throughout the Outer Lands. Phosphorous derived from laundry soaps and home detergents also leaks

from septic systems into groundwater and local ponds and streams, where it helps drive eutrophication. In addition to hypoxia, excess nitrogen in marine systems can destroy an important coastal environment—the Eelgrass community. On the Cape and Islands, in Peconic Bay between the north and south forks of Long Island, and in Long Island's Great South Bay excess nitrogen has destroyed the once-rich Eelgrass beds, and with the loss of Eelgrass the young of many important food and sport fish no longer have the food and shelter they need to survive to adulthood in more open coastal waters. The Eelgrass habitat losses in Buzzards Bay and Waquoit Bay are particularly well documented, and both are tightly related to excess nitrogen in coastal waters.

Further declines in commercial fishing

Anyone who has visited the Outer Lands regularly over the past few decades has seen the sharp drop in commercial fishing operations. Traditional New England commercial fishing centers such as Point Judith, Provincetown, and Chatham that were bustling with activity in the 1970s and 1980s now host fewer and fewer boats and crews.

The Magnuson-Stevens Fishery Conservation and Management Act of 1976 established a 200-mile jurisdiction zone in US coastal waters and for the first time instituted fishery-wide management principles for the continental shelf waters and fishing banks within US waters, such as Georges Bank. The Magnuson-Stevens Act also excluded foreign factory ships from US coastal waters, and this led to a major expansion in

Commercial fishing boats anchored near the Chatham Pier, Cape Cod. Today Chatham's fleet is only a fraction of the size it was before falling fish populations made commercial fishing less viable. Storm breaches in Chatham's barrier islands in 1987 and again in 2017 have also made the harbor more vulnerable to ocean waves and shifting sandbars.

the East Coast commercial fishing fleet. Although the federal government instituted many controls on the equipment and catch sizes of target species, the controls proved to be ineffective as one species after another became too rare to support a fishery in the Northwestern Atlantic.

The issues around commercial fishing regulation are complex and well beyond the scope of this guide, but a few factors are likely to guide the future of commercial fishing in the Outer Lands and Western Atlantic Ocean.

• Nature doesn't always recover once we protect a target species. The Atlantic Cod has been heavily protected since the Northwestern Atlantic populations crashed in the 1990s, but the species has yet to show significant signs of recovery even 25 years later. It seems that there are tipping points beyond which a fish population cannot recover, at least not in human time frames. We have little reason to believe that now-depleted species such as Redfish, Atlantic Halibut, Yellowtail Flounder, and Haddock will do any better than the cod did, and substantial recovery in these species may not occur in the lifetimes of today's commercial fishers.

• With our modern fish-finding and catching technologies we are capable of destroying species fisheries very quickly, unless we have strict regulation on annual catch sizes of target species, with solid scientific information on the current status of each fish species.

• Despite decades of increasingly strict regulation and failing fish populations, we still have too many commercial fishers chasing too few fish.

Commercial fishing boats, Chatham Harbor.

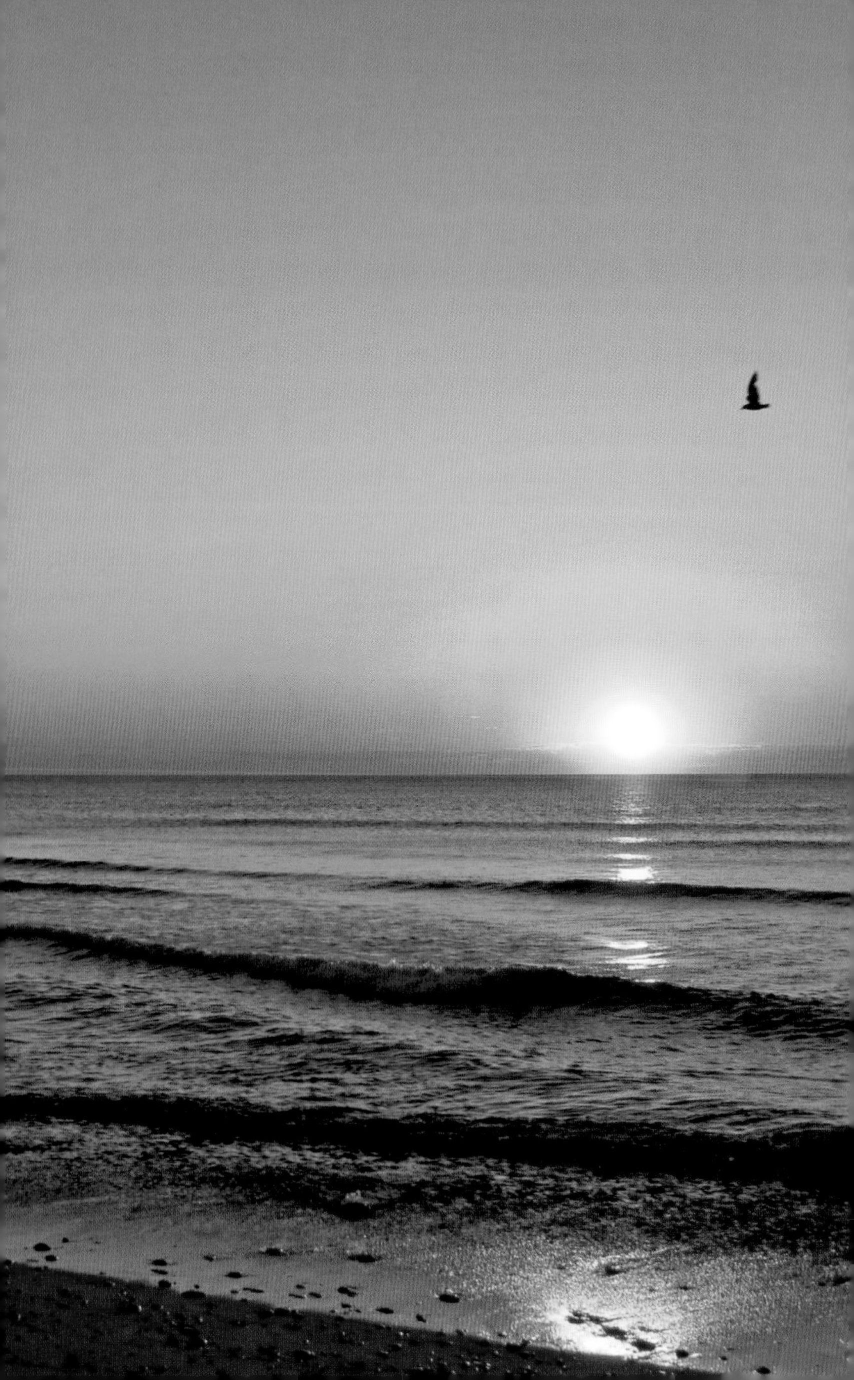

BEACHES

The earthen marine scarps above Coast Guard Beach in Eastham erode easily, about three feet per year.

Beaches are rough and turbulent places, built and eroded by winds, waves, and coastal currents. For plants, beaches are dry as deserts and sprayed with blasts of sand and salt. Beaches respond with great sensitivity to all these forces, and each ripple line or small ridge of sand reflects those shaping energies like a sculptor's handprint. Everything on a beach is in constant motion, whether under the blazing sun of summer or the grinding ice and tearing waves of winter. The ultimate fate of the beach lies in the thin swirl of sand within each wave swash. Are the waves bringing new sand to build the beach or tearing away the substance of the strand?

Time and erosion are the mightiest factors in shaping our beaches but the hardest to visualize. The Great Beach of the Outer Cape seems timeless—Thoreau would recognize it instantly even today—yet because of erosion, the Outer Cape Cod beaches that Thoreau walked in the 1840s were more than 500 feet offshore of today's beaches. Eastham's Coast Guard Beach and Nauset sandspit may be the Cape's loveliest stretches of shoreline, but the beach Henry Beston wrote of in *The Outermost House* was all but destroyed in a 1978 nor'easter, as was his house and the dune ridge it once sat upon. The beach Beston walked in the mid-1920s was about 275 feet seaward of today's much more modest Nauset sandspit. If, as Heraclitus said, "No man ever steps in the same river twice," the same is true for any beach. The beach you walk today is not the beach of last week or last year. Even if the place seems the same, every grain of sand you see was somewhere else just a short time ago.

Opposite:
Dawn at Lecount Hollow Beach, Wellfleet.

Opposite:
Unlike the river-borne sands of beaches south of New York Harbor, all Outer Lands beaches are made up of sand from eroding glacial till along the coast. Here the 100-foot cliffs above Coast Guard Beach in Truro show fans of eroded sand and silt at their base. On average these cliffs erode back about three feet per year, with most of the loss during winter storms. Note the rectangular concrete foundations on the cliff rim, at the top left. The foundations were originally built far back from the cliff edge, but now they are almost ready to tumble down the cliff face.

The sands of the Outer Lands aren't brown because they are dirty. The sand appears brown because it's a relatively young blend of many minerals. Viewed close up (see inset), the sand is a mix of brightly colored grains. When you blend the colors in your mind's eye, you see brown.

The structure and formation of beaches

The constant action of waves and currents sorts rocks and sand by size, forming large areas of consistent sediments. Water sorts sediments into coarse pebble areas, finer sand zones, and finest of all, silty clay sediments. The fine particles of silt move the farthest and can drift in the water column for years before settling either deep offshore in the ocean bottom or as muddy sediments in bays and other protected areas with slower, more gentle wave action. Sand accumulates along shores because it is composed of such hard granitic minerals as quartz and feldspar in particles small enough to be easily moved by ocean waves but also durable enough to persist along the coastline rather than be washed immediately into deep water.

All Outer Lands beaches ultimately originate from mixed glacial till torn by waves from the soft earthen shore edges and cliffs that are so common in this region. By contrast, most Atlantic Coast beaches from New York Harbor south to Florida's northern border are created by sand eroded from upland areas and ferried to the coast by major rivers. The Cape and Islands and Long Island have no rivers large enough to move significant quantities of sand, so all the sand you see, even in such large, sandy areas as Cape Cod's Provincelands, Monomoy Island, and Sandy Neck or on New York's Fire Island, is derived solely from erosion of Outer Lands coasts.

The Outer Lands beaches consist chiefly of sand and finer gravels from eroded glacial till that is quite young in geologi-

The earthen cliffs (marine scarps) above Coast Guard Beach, Truro.

A huge cliff collapse, Coast Guard Beach, Truro. Most erosion of the marine scarps near beaches happens during winter storms or in spring, when the combination of storm damage and spring rains weakens the earthen cliffs. In April 2016 a two-acre section of the 100-foot cliffs above Coast Guard Beach suddenly collapsed onto the beach, spilling huge chunks of sandy clay and bank vegetation down to the high tide line. Although collapses of this size are rare in beach season, please pay attention to the signs warning beachgoers not to climb on or set beach chairs too close to the cliff faces.

cal terms and has not been exposed to the long polishing and grinding that create the fine, white sugar sand of the prettiest beaches to the south. These young, rough sands are mostly quartz, with fragments and sand-sized grains of other softer and mostly darker rocks and minerals. In time the softer minerals will be ground into silt and washed away, but since this sand was created largely by glacial events less than 35,000 years ago, these darker minerals remain, giving the beaches a darker look than the beaches of the unglaciated Atlantic Coast south of New York Harbor, where much of the sand is far older. In addition, Cape Cod Bay's lower-energy beaches sometimes have muddy areas of silt-sand mixtures because the currents and waves aren't always strong enough to wash away the silt particles, as would happen on an ocean beach.

Most sand on Outer Lands beaches is light brown, but streaks of darker or colored sand often appear, particularly in wet sand near the swash zone. "Oil pollution" is often blamed for dark patches of sand, but the cause is almost always natural. The dark streaks are heavier minerals such as magnetite. The swash water sorts the lighter quartz and feldspar sand grains from the heavier magnetite, leaving dark streaks. In a similar way, red or purple streaks may be caused by garnet grains, and green streaks by grains of the mineral hornblende.

One common kind of black beach stone tells a sadder story. On the outer ocean beaches of Cape Cod you may find rounded, sand-smoothed lumps of coal. When coal was commonly used for home heating, great four- and five-masted wooden sailing ships transported coal up and down the East Coast. As the heavy ships rounded Cape Cod they were sometimes wrecked on the outer bars by storms. Storm waves have long since destroyed the ships, but their loads of coal remain offshore, and occasionally pieces come ashore, like small memorials to their vanished ships and crews.

Wind and sand

On the upper beach, just above the splash of waves, sand grains begin to move under the influence of winds. At low tide the sand dries, and when winds reach a speed of about 12–15 miles per hour, they can move average-sized sand grains across the beach.

Several processes transport sand grains. Strong winds pick up surface grains and transport them in short leaps across the surface in an action called saltation (from the Latin for "jumping"). Wind also can shove the grains of sand directly along the beach surface in a process called surface creep. As any beachcomber can tell you, wind is quite capable of moving sand grains well above the surface—to eye level at least. Any windy day will move great quantities of sand, but most beachgoers never see the strong, steady winter winds and blustery storms that move the majority of sand over the year.

Onshore and offshore winds

Beaches are windy for the same reason that open waters are windy: there are no landforms or tall vegetation to break the force of the wind. Beaches are also windy because of the

These dark streaks across the sand at Marconi Beach in Wellfleet are composed of magnetite, a common dark-colored mineral. The waves sort the heavy magnetite grains from the lighter quartz and feldspar grains, leaving streaks of black. The dark violet hues are from the mineral garnet.

2 inches		Paths of sand grains	
1 Wind		Each grain makes many rebound leaps across the sand surface	
Surface creep: Wind can simply push or roll sand grains along the surface	Saltation: Wind picks up sand grains and transports them short distances		Vegetation is crucial to help build up dunes and barrier islands by trapping sand grains carried by the wind

Marconi Beach, Cape Cod National Seashore, on a blustery day in April,

Natural beaches, such as Coast Guard Beach in Truro, are in constant flux, changing in size and vertical profile with the seasons. Beaches have enormous aesthetic and recreational value and help protect the coast by absorbing the destructive wave energy of storms. When we insist that the beach cannot change and build hard structures too close to the tide line, we court disaster.

very different ways that land and sea heat and cool over the seasons. This differential heating and cooling along the coast generates onshore winds as well as their opposite, offshore winds, particularly in summer, when the contrast between land and sea temperatures is the greatest.

Onshore winds are generated when the land heats up during the day, creating columns of rising warm air. As the warm air mass rises, it draws in air from the lower, cooler air over the sea, creating winds that blow in from the sea, usually in the afternoon and early evening. While these cool sea breezes may feel great on land, the onshore winds can be strong enough to kick up three-to-four-foot waves even in sheltered waters, sometimes creating a rough ride for small boats. Onshore winds happen almost daily along Outer Lands coastlines in summer and early fall.

Offshore winds result when the ocean is warmer than the land and rising air over the ocean pulls air from the land. Offshore breezes most often occur in late fall, when the water is still relatively warm but the land is cooling quickly as winter approaches.

Groins, jetties, and breakwaters

In the past, various kinds of hard structures have been built to try to stabilize sand beaches and protect them from storm damage. Breakwaters are long rock structures built offshore and roughly parallel to the shoreline to reflect storm waves and protect harbors. The long, bird-covered line of rocks just off MacMillan Pier in Provincetown Harbor is a classic breakwater. Jetties are walls built on either side of a harbor entrance to prevent the navigation channel from filling with sediment carried by longshore currents and to stop the development of a bay-mouth sandbar. The long lines of rock that extend from

Jetties Beach and Coatue Point on either side of the entrance to Nantucket Harbor are jetties that prevent the channel from filling with sand from the longshore drift currents. Groins are stone walls built perpendicular to sandy shores to stabilize and widen beaches that are losing sand to erosion from long-shore currents. Groins are no longer considered effective for shoreline engineering, because although they trap some sand locally along the upstream face of the wall, they rob down-stream areas of their sand supply. Over time, groins worsen the overall problem of beach erosion.

How to read a beach

Although all beaches share superficial similarities, each beach has differences in wave energy, sand origin and supply, and weather regime that together determine its unique shape and character.

Sandbars and the waves they generate

If you look carefully at the waves approaching an ocean beach on a day with moderate weather, you'll see a number of

Hardened coasts destroy beaches because they reflect most wave energy rather than absorbing it. The reflected wave energy from the rocks tears away beach sand. Here in eastern Provincetown the reinforced coast along Commercial Street has destroyed most of the original beach, which is now visible only at low tide (shown here). Note the two groins along the coast at the middle right of the photo. The upstream groins have trapped some sand, at the expense of the beach in the foreground.

A breakwater protects the inner harbor and piers at Provincetown.

general patterns in the position and timing of breaking waves, as well as in the shape of the breakers themselves. The first pattern to notice is determined by the position of sandbars. Two sandbars form at most ocean beaches. The nearshore bar forms close to the beach, just seaward of the wave breaks nearest the shoreline. The offshore bar is marked by a second line of breakers. Offshore bars are very common on beaches exposed to strong ocean waves, such as the beaches along Cape Cod National Seashore from Chatham north to Race Point and on Long Island from Montauk Point to Jones Beach. On exposed ocean beaches with large waves the offshore bar may be 100–200 yards from shore, particularly in winter, but on most beaches and in summer the offshore bar is much closer to shore. The notorious ship-wrecking Peaked Hill Bar of the Outer Cape is a large offshore sandbar that begins at Head of the Meadow Beach in Truro and extends northwestward around the curve of the Provincelands almost to Race Point (see illustration, p. 48). In the days of sail the Peaked Hill Bar claimed hundreds of fishing vessels and cargo ships that were caught in northeasterly storm winds and pushed onto the bar, where they grounded and were then torn to pieces by the large wave sets breaking over the bar. Most fishermen and sailors could not swim, and they drowned within easy sight of the beach.

As ocean waves approach a beach, they begin to slow and the spacing between waves decreases, causing the wave shapes to steepen, and finally the steep structures topple forward and break. As with sandbars, the position of wave breaks can tell you a lot about the beach profile under the water, and the pat-

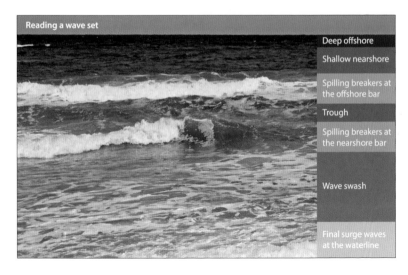

Reading a wave set

Deep offshore

Shallow nearshore

Spilling breakers at the offshore bar

Trough

Spilling breakers at the nearshore bar

Wave swash

Final surge waves at the waterline

Spilling breakers

Surging breakers

Plunging breakers

tern of breaking waves can also reveal the overall wave energy as they interact with the bottom profile. Most smaller ocean waves approaching a beach have a period (time in seconds be-tween wave crests) of about 8–15 seconds, with larger storm waves showing a longer 15–20-second interval.

Types of wave breaks

Wave breaks fall into three general patterns. Most fair-weather ocean waves are medium-energy waves known as spilling breakers. As each breaker approaches the shallows, its steepening crest curls just a bit as it spills down the front, tak-ing a relatively long time to break fully. Spilling breakers seem to simply fall apart as they hit the shallows. Surging breakers are another form of fair-weather wave, often seen passing over the offshore sandbar without breaking. The long-period, low-profile waves push quickly over the shallows without much white water, then hit the wave swash from the previous waves and disintegrate into a low surge of white water, again without much force. Plunging breakers are the classic high-energy storm- or wind-driven waves that heave into a near-vertical wall as they approach the shallows and then form the classic tubular curve that surfers love and most people fear. Plung-ing breakers are usually the result of offshore storms, which

The slope angle of a beach also has a major effect on the size and nature of breaking waves. In summer the broader, less steep beaches mostly have spilling breakers. The steeper, narrower beaches of winter cause more waves to break in the classic curl.

can send long-period waves across great distances of ocean. Plunging breakers hit the shallows with tremendous force, sometimes pounding the beach so hard that you can feel the break through your feet, even if you are many yards away up the beach. A major storm or very strong onshore winds can cause a storm surge, creating a dangerous wave pattern called a shore break, in which large waves pass over the sandbars without breaking (because of the deeper surge waters) and break directly onto the beach. Swimming in a stormy shore break is dangerous: the large waves breaking just off the beach can sweep you up and pound you directly into the sand with enough force to cause injury.

Rip currents

If you swim on ocean beaches, it is critical to understand rip currents: what they are, how they form, and, most important, what to do if you are caught in one. Rip currents (sometimes incorrectly called rip tides) form when strong sets of waves drive more water up a beach than can easily drain away through the normal swash of waves. As each succeeding wave breaks and slides down the beach, the next wave blocks the backflow of water and raises the overall level of water on the beach. Eventually too much water is well above the current tide level. To release the pent-up water, a swift current forms in a low spot or channel in the beach and drains the excess water out to sea. This channel of fast-flowing water, normally running perpendicular to the shoreline, is a rip current (see

Rip currents are most common on the Outer Lands beaches that are most exposed to large ocean waves, such as the ocean beaches along the Outer Cape and the beaches on the south shore of Long Island.

Rip currents
Watch for odd gaps in the normal pattern of incoming waves. Often rips are turbulent with much white water. Other times they are dark because of the deeper water channel. This is a rip current in a drainage area at Marconi Beach.

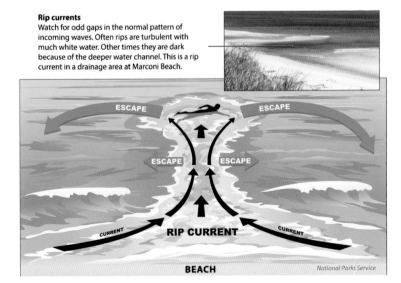

National Parks Service

diagram, p. 120). Rip currents are powerful and can flow at three to five miles per hour. Even an Olympic-caliber swimmer could not swim against a current that strong. If you are caught in a rip current, swim parallel to the beach to exit the main flow, as most rip currents are narrow channels.

Many rip currents are visible if you know what to look for. Watch for a dark or light gap in the normal wave sets coming onto a beach. Sometimes a strong rip current kicks up a lot of white water. Rip currents are often a different color than the water around them because they run in a deeper channel or because they are drawing silt and sand off the beach.

Beach profiles

The actions of waves on sandy beaches over the seasons produce a consistent and predictable beach structure. The foredune is a raised portion at the back of a beach that is a transition point from true beach into a vegetated dune. The foredune is raised because the beach plants there are dense enough to trap and hold sand, and over time the foredune rises over the height of the upper beach. The upper beach is created largely by wind action and is a repository for sand grains blown up from lower areas. The upper beach is usually bare, though it may harbor a few very hardy plants, such as American Beachgrass, Dusty Miller, Seaside Goldenrod, Beach Pea, Beach Clotbur (Cockleburr), and Common Saltwort.

Wrack lines

The wrack line (also called the wet-dry line) marks the average high tide line. Incoming waves are more forceful than the backwash off the beach, and as they come in, the waves

On ocean beaches the upper beach is often marked by a distinct step in the beach called the winter berm, created when large (usually winter) storm waves reach high up the beach and pull sand from areas that are normally well above the usual high tide line. Here at Marconi Beach in Wellfleet winter storms created a four-foot winter berm on the upper beach in March.

This photo also illustrates how different the summer beach profile (upper sand level) is from the lower winter sand level below the berm.

wolf spider
Pardosa sp.

sweep material from the tidal zone and onto the beach up to the maximum high tide line. This leaves a collection of plant and animal debris called the wrack line. On relatively sheltered coasts such as those on Cape Cod Bay or Nantucket Sound, wrack lines are usually composed of broken stalks of *Phragmites* (Common Reed) and Saltwater Cordgrass, strands of Eelgrass, molted crab shells, Sea Lettuce and other algae, and bits of other floatable debris. In the mix may be clam and scallop shells, egg cases from whelks and skates, and the carcasses of various small fish and invertebrates. Most crab shells you see in the wrack line are not from dead crabs but are the molted shells that crabs shed as they grow.

Wrack lines are inhabited by beach flies, wolf and other spiders, beach fleas (amphipods), and beetles, all of which are important sources of food for shorebirds that feed on the beach. Most wrack line animals are nocturnal, feeding at night when the beach is cooler and damper. Wrack line shells and plants are a good indication of what plants and animals are abundant just offshore in the subtidal environment. Dead and partially dismembered Northern Searobins are common in the wrack line because they are easily caught by fishers and make good bait for Bluefish and Striped Bass.

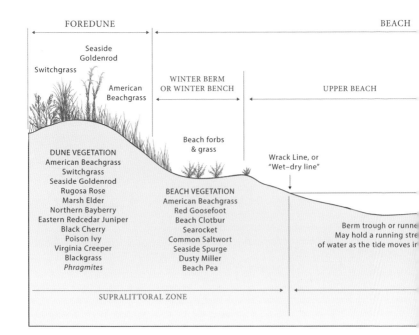

FOREDUNE

BEACH

Seaside Goldenrod

Switchgrass

American Beachgrass

WINTER BERM OR WINTER BENCH

UPPER BEACH

Beach forbs & grass

DUNE VEGETATION
American Beachgrass
Switchgrass
Seaside Goldenrod
Rugosa Rose
Marsh Elder
Northern Bayberry
Eastern Redcedar Juniper
Black Cherry
Poison Ivy
Virginia Creeper
Blackgrass
Phragmites

BEACH VEGETATION
American Beachgrass
Red Goosefoot
Beach Clotbur
Searocket
Common Saltwort
Seaside Spurge
Dusty Miller
Beach Pea

Wrack Line, or "Wet–dry line"

Berm trough or runne
May hold a running stre
of water as the tide moves in

SUPRALITTORAL ZONE

The seeds of many annual beach plants are also mixed into the wrack line, and their dispersal by tides, wind, and wave action are an important way that these plants spread, particularly on more protected beaches away from heavy ocean waves. Seaside Spurge, Common Saltwort, Red Goosefoot (Coast-Blite), and Searocket are annuals that spread their seeds this way. Wrack debris aids a beach by trapping and holding windblown sand, contributing to dune formation. Many foredunes begin as sand and seeds trapped in a high storm wrack line that gradually accumulates enough sand to support plant life.

Many bathing beaches sweep away the wrack line to leave "clean" sand. Unfortunately, beach grooming for sunbathers tends to sterilize the beach, removing food sources, shelter, and seeds for plants and animals. Ironically, it also makes the sand more mobile and thus more prone to being blown away or swept away by tides. Grooming is particularly bad for two of the Outer Lands' most endangered birds, the Piping Plover and the Least Tern, both of which nest on beaches. The Piping Plover also feeds frequently in the wrack line where it nests, and beach grooming often destroys Piping Plover nests.

Wrack lines provide food and shelter for many small beach organisms. Beach fleas (amphipods), small crabs, wolf spiders, many other kinds of spiders, beetles, and even foxes and Raccoons scavenge in wrack lines. Wrack lines also trap and hold windblown sand on beaches.

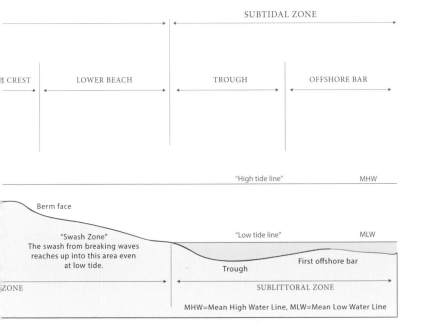

SUBTIDAL ZONE

CREST | LOWER BEACH | TROUGH | OFFSHORE BAR

"High tide line"   MHW

Berm face

"Swash Zone"
The swash from breaking waves reaches up into this area even at low tide.

"Low tide line"   MLW

First offshore bar

Trough

ZONE   SUBLITTORAL ZONE

MHW=Mean High Water Line, MLW=Mean Low Water Line

seaweed fly
Fucellia sp.

Small wrack line invertebrates such as seaweed and beach flies (*Fucellia sp.*) play a crucial role in the beach food chain, converting detritus, bacteria, small microorganisms, and plant debris into animal matter. The flies are important food sources for migrating shorebirds, as well as resident songbirds and other dune birds nesting near the beach.

The lower beach

The lower beach or swash zone is the transition area where waves meet the beach, swashing up the sand and sliding back with each new wave set. Farther up the beach a summer berm typically forms where the relatively gentle summer waves pile up an accumulation of sand, building the width and height of the beach. Often the beach becomes steeper on the seaward side of the summer berm.

Just below the foreshore area of beaches, below the low tide line, there is typically a trough in the beach profile where heavier gravel and shells accumulate, and just beyond that dip area is a shallow bar area composed of a mix of the coarse materials and sand brought in from deeper waters. It seems

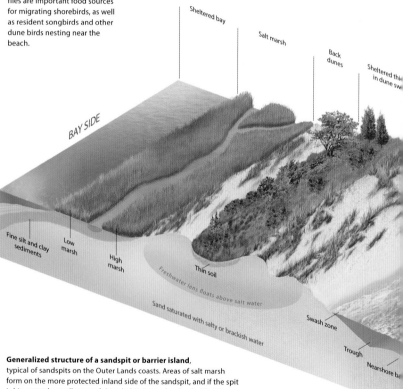

Sheltered bay

Salt marsh

Back dunes

Sheltered thi in dune sw

BAY SIDE

Fine silt and clay sediments

Low marsh

High marsh

Thin soil

Freshwater lens floats above salt water

Sand saturated with salty or brackish water

Swash zone

Trough

Nearshore ba

Generalized structure of a sandspit or barrier island, typical of sandspits on the Outer Lands coasts. Areas of salt marsh form on the more protected inland side of the sandspit, and if the spit is big enough, small areas of dune habitat may develop behind the foredune and beach.

that every child at the beach makes the surprise discovery that if you head out into the water and brave the first line of breakers and the deep trough under them, you'll suddenly be in much shallower water when you reach the sandbar beyond the breaking waves. Offshore sandbars are formed when sand pulled off the lower beach by wave swash is drawn into deeper water just offshore, where wave action below the surface is less intense. The sand milling in the backwash drops out of the water column, forming a sandbar just seaward of the beach trough.

Beach plants

Few plants and animals are well adapted to the dry conditions, sandy substrate, and salt spray of beach environments. No plants can survive below the wrack line except Saltwater Cordgrass on sheltered beaches, and few plants can survive the first stretch of upper beach behind the wrack line, primarily because only a few plants can tolerate having their roots immersed in salt water. Most true beach plants live in the upper beach area just before the rise of the foredunes behind the beach, where American Beachgrass begins to dominate.

Saltwater Cordgrass (*Spartina alterniflora*) is a common salt marsh grass that also grows at the water line on more sheltered beaches.

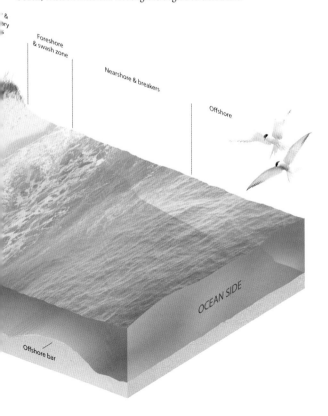

Foreshore & swash zone

Nearshore & breakers

Offshore

OCEAN SIDE

Offshore bar

Red Goosefoot or Coast-Blite (*Chenopodium rubrum*) is one of the few beach plants that grows right up to the wrack line.

American Beachgrass (*Ammophila breviligulata*) is the most important shoreline plant north of Cape Hatteras in stabilizing coastal dunes, sandspits, and barrier islands.

American Beachgrass

American Beachgrass is the most common plant on the upper beach, and it becomes even more dominant in the foredune and back dunes areas beyond the upper beach. A number of features allow it to thrive under sandy, dry conditions. Its leaves curl into vertical tubes to conserve moisture in dry winds. Beachgrass can spread by seed, but it generally spreads over a beach through underground stems or rhizomes, with roots that extend deep into the sand—six feet or more is not unusual. Because of its rhizomes and deep roots safe under the sand, American Beachgrass can survive storm damage, sand blowouts, and winter exposure.

American Beachgrass is often the first plant to colonize an empty stretch of beach or dune. Once its seeds take root, the grass spreads rapidly through its rhizomes, and as the grass spreads, it begins to build sand dunes or sandspits by accumulating windblown sand particles. As the leaves trap the sand grains, a small mound forms around each grass cluster. One of this grass's most important adaptations is the ability to grow upward quickly, which keeps it from being overwhelmed by the growing pile of sand around it. Many other beach plants have similar vertical growth adaptations for the same reasons. As American Beachgrass moves into a beach or sandbar, the grass literally builds its own environment by accumulating and stabilizing the sand with its roots and rhizomes. The grass does not just grow on sandspits and dunes—American Beachgrass creates the sand dunes of the upper beach.

Other common beach plants

Saltwater Cordgrass is one of the few land grasses that can tolerate having its roots soaked in salt water, and on protected beaches without much wave action you'll often see small stands of Saltwater Cordgrass running right down the beach to at least the low tide line and sometimes a little beyond.

In late summer and fall Seabeach Orach lines upper beaches, gradually turning red as the weather cools. Seabeach Orach seeds are highly salt tolerant, and they overwinter in the wrack line to start a new generation the following spring.

Dusty Miller, Beach Clotbur (Cockleburr), Searocket, Common Saltwort, and Seaside Spurge are all species adapted to living on beaches and dune faces. They have tough, leathery leaves that are sometimes also waxy or hairy, all strategies to prevent moisture loss. Seaside Spurge grows in a flat disc just above the level of the sand to conserve water by staying out of the wind. If you look closely at Seaside Spurge plants in late summer or fall, they seem to grow on top of little mounds. As with American Beachgrass, the spurge creates the mound by trapping sand between its leaves and stems and thus must constantly grow upward or be buried under the mound of sand captured by its own leaves.

Seaside Goldenrod is one of the most common beach specialists, and seems to dominate the upper beaches in the late summer and fall with its showy yellow sprays of flowers. This hardy perennial goldenrod has tough, waxy evergreen leaves over a deep set of roots that often stretch two feet or more underground. Seaside Goldenrod seems particularly resistant to salt spray damage and often grows right up to the edges of upper beaches.

Beach Clotbur or Cockleburr (*Xanthium strumarium*) seeds are a common sight on Outer Lands beaches in summer and fall.

Two of the Outer Lands' most endangered bird species nest in the upper beach zone: the **Piping Plover** and the **Least Tern**.

Dusty Miller (*Artemisia stelleriana*) is very common on the upper beach and the first sand dune lines behind beaches.

Rugosa Rose (*Rosa rugosa*) often marks the inner edge of the upper beach and the transition into a dune environment.

Farther back on the upper beach, in the transition to dune habitats, Rugosa (Salt-Spray) Rose and Beach Pea become common. Rugosa Rose is an Asian exotic introduced to the Atlantic Coast in the 1800s for its ability to grow in and stabilize sand dunes. Rugosa Rose is the most salt spray–tolerant shrub and is often the first shrub to appear in the upper beach area.

Beach animals

Most beach animals are not visible to the casual observer because they live within the sand and rarely come to the surface before nightfall. Much of the subsurface beach life is also too small to be viewed without magnifiers or microscopes. These tiny animals, collectively called the meiofauna, live between the grains of sand in the lower beach area that stays moist between tides. Mites, ostracods, tardigrades, copepods, nematodes, and various worms all swarm out of sight within an inch or so of the sand surface, and they are a valuable food supply for the larger copepods, shrimp, and beach fleas that in turn feed the shorebirds. The beach meiofauna have only recently been discovered and studied, and little is known about the role they play in the larger ecology of beaches.

Sponges are common subtidal animals that most of us see only when they wash up on beaches. The Red Beard Sponge is a recent invasive species often seen washed up on beaches. Fresh specimens are bright red, and even when dead and dried on the beach, the branching fingers of Red Beard Sponge retain a faded but distinct red color.

The swash zone of the lower beach may appear lifeless to the casual observer, but if you look closely or watch shorebirds like Sanderlings work the swash between waves, there is more there than you might suppose. One of the best-adapted residents of the wet beach is the tiny, egg-shaped Atlantic Sand Crab, often called the Mole Crab. As its name suggests, the Mole Crab lives just under the surface of the wet sand at the wave's edge, with only its antennae and part of the head exposed just above the sand. Mole Crabs use their comblike antennae to sweep plankton and tiny animals out of the water as the waves swash over them. Most ocean beaches and many bay beaches have Mole Crabs. As you stand in the swash zone, try scooping up a large handful of wet sand and watch the exposed crabs swiftly rebury themselves in the sand you hold.

The American Horseshoe Crab is a large but completely harmless creature often spotted on beaches, unfortunately not always alive, because these crabs have been overharvested as fishing bait. Only recently, as their numbers have dwindled, have they attracted conservation support and research interest. In May and June, often during a spring (unusually high) tide, Horseshoe Crabs travel up into the low tidal zone to lay their eggs. These eggs are very attractive to shorebirds, and often the best indication that the crabs are breeding is the sight of flocks of birds avidly picking at the eggs in the surf line.

Northern Moon Snails and Atlantic Oyster Drills are predators on bivalves in the intertidal region of beaches. Both species drill neat, round holes in shells, usually near the umbo, or hinge point, of a clam or oyster. Once the snail finds a suitable live clam, it uses its sharp, toothy radula to scrape a hole in the clam's shell as it secretes a strong acid to help dissolve the shell. When the radula breaks through to the interior of the clam shell, the snail injects powerful enzymes that digest the clam's interior organs and muscle. The snail then sucks up the dissolved clam through the hole in the shell.

Clam or oyster shells that you find on beaches are often riddled with holes from snails but may also be eaten away into a honeycomb of holes through the action of Boring Sponges. Boring Sponges don't prey on clams and oysters directly, but the slow dissolving of their shells is usually fatal to oysters if the sponge attacks an occupied shell.

The shells and shell fragments you see on the beach are partly from clams and snails that have recently died, but many of the shells you see are hundreds of years old. If a shell is buried under sand on the beach or offshore, the shell may stay intact for a very long time before it is reexposed to air. Dark brown or even black shells may have been buried in mud or marsh peat for decades or even centuries. Researchers collected beach shells on Plum Island, Massachusetts, ground up the shells to form a representative mixture for age analysis, and found that the average age of the shells was 400 years. Have some respect for those old clam shells!

Dead or dying jellyfish often appear in the swash zone just below the wrack line. Moon Jellies are usually pale greenish or bluish puddles of gel about six inches in diameter. The larger Lion's Mane Jellyfish is red-brown in color with patches of lighter colors. Atlantic Sea Nettles are usually smaller than Moon Jellies but have long, trailing tentacles. Beware of approaching a stranded jellyfish if you are barefoot. The stinging tentacles remain viable long after the animal has died, and the tentacles can spread several feet from the remains.

Tidal flats
Tidal flats and nearby subtidal areas with sandy or muddy bottoms have a rich infauna of animals that burrow into the bottom sediments for shelter. Most clams and marine worms,

Back side

Under side

Atlantic Sand Crab (Mole Crab)
Emerita talpoida

Seaside Goldenrod (*Solidago sempervirens*) is one of the most common and noticeable beach plants all along the northern and central Atlantic Coast.

AMERICAN BEACHGRASS *Ammophila breviligulata*

SEASIDE GOLDENROD *Solidago sempervire*

BEACH CLOTBUR *Xanthium strumarium*

SALTWATER CORDGRASS *Spartina alternifl*

COMMON SALTWORT *Salsola kali*

COMMON SALTWORT, detail

SIDE SPURGE *Chamaesyce polygonifolia*

SEASIDE SPURGE, detail

GOOSEFOOT *Chenopodium rubrum*

RED GOOSEFOOT, detail

ROCKET *Cakile edentula*

SEAROCKET, detail

SANDBUR *Cenchrus longispinus*

YARROW *Achillea millefolium*

EASTERN PRICKLY-PEAR *Opuntia humifusa*

DUSTY MILLER *Artemisia stelleriana*

BEACH PEA *Lathyrus japonicus*

SEABEACH ORACH *Atriplex pentandra*

SON IVY *Toxicodendron radicans*

VIRGINIA CREEPER *Parthenocissus quinquefolia*

BRIER *Smilax glauca*

BLACK SWALLOW-WORT *Cynanchum louiseae*

Weges

NEBERRY *Rubrus phoenicolasius*

JAPANESE HONEYSUCKLE *Lonicera japonica*

Northern Moon Snail
Euspira heros

Slipper Shell
Crepidula fornicata

Atlantic Oyster Drill
Urosalpinx cinerea

Northern Quahog
Mercenaria mercenaria

some crabs, and even some fish bury themselves at least partially as protection in flat bottom areas that often lack rock or plant shelter. The beaches along the shores of Cape Cod Bay are famous for their shallow tidal flats that can extend out from the beach for miles at low tide. The John Wing Trail that runs north of the Cape Cod Museum of Natural History leads to a beautiful wild beach and tidal flat that is particularly good for exploring this environment. If you venture out onto the tidal flats, be absolutely sure that you understand the tide schedule and don't get caught having to wade back to shore.

Tidal flat invertebrates

Soft sediment bottoms and tidal flats can look deceptively lifeless unless you look for the single or paired siphons of clams buried within them. Northern Quahogs have short, paired siphons, and the top edge of their shells is rarely buried more than an inch below the surface. In smooth sand or mud bottoms, look for a figure eight of the twin open siphon holes. Quahogs prefer a salinity of around 20 ppt (parts per thousand) or greater, so they are less common in the soft bottoms of river mouths. Quahogs are called by a variety of names based on their size, but little necks, cherrystones, chowder clams, and quahogs are all the same species: the Northern Quahog. Softshell Clams, or steamers, have an extremely long, tough pair of siphons encased in a thick black membrane. The long siphons allow the Softshell Clams to bury themselves far below other clams, sometimes 10 inches deep. The Atlantic Jackknife Clam, or razor clam, has a very short siphon that looks keyhole-shaped at the surface. These clams bury themselves vertically with a short but strong foot on the lower end opposite the siphon. Atlantic Jackknife Clams sometimes pop up above the surface, often when disturbed by mud worms probing their burrows from below. If a Jackknife Clam senses movement, it disappears in a flash into its burrow.

The largest bivalve in the Outer Lands, the Atlantic Surf Clam, lives in sandy bottoms from the subtidal zone down to several hundred feet. This clam is surprisingly long-lived, living for 31 or more years. Most surf clams are harvested at about 15–20 years, primarily for chowders and fried clams. Their chief predators are Moon Snails, Horseshoe Crabs, Atlantic Cod, and, of course, humans.

One of the most common shelled animals of the subtidal zone is a sea snail, the Slipper Shell, often called a boat shell. These snails are common in both soft sand and rocky, shallow subtidal areas and on flats and beaches exposed at low tide. They are filter feeders and typically live in stacks, with older individuals at the bottom and successive layers of younger individuals attaching on top of the older shells. A muscular foot holds each individual in place in the stack, and when submerged, the foot relaxes slightly to open a gap through which the snail draws water to filter for plankton.

Beach birds

The dominant birds of Outer Lands beaches are gulls, particularly the Herring Gull, Ring-Billed Gull, Laughing Gull, and Great Black-Backed Gull, joined in the winter months by the more unusual white-winged Iceland and Glaucous Gulls. Gulls are intelligent and watchful predators of all small beach animals (including nestling birds of many species), taking their prey from the immediate shoreline or from the shallows near the beach. Their usual prey includes clams, crabs, snails, small fish, and any other animals they can capture or scavenge. Gulls generally do not dive for their prey, which limits their feeding to the shallowest water and the beach and tidal flat areas exposed at low tide.

The gull population of the Outer Lands region expanded rapidly in the mid-twentieth century with the rise of suburban towns using open-air dumps and large-scale commercial fishing offshore. However, in the past decade gull populations have contracted substantially, as open-air dumping has become much less common, factory fishing has vanished, and

Softshell Clam
Mya arenaria

Invertebrates in and near the bottom sediments of the subtidal zone.

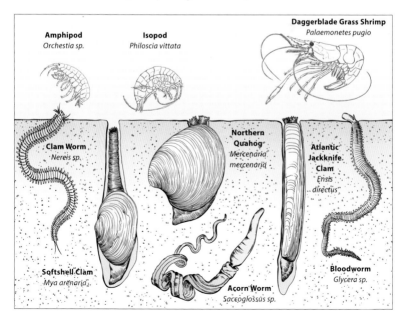

Amphipod
Orchestia sp.

Isopod
Philoscia vittata

Daggerblade Grass Shrimp
Palaemonetes pugio

Clam Worm
Nereis sp.

Northern Quahog
Mercenaria mercenaria

Atlantic Jackknife Clam
Ensis directus

Softshell Clam
Mya arenaria

Acorn Worm
Saccoglossus sp.

Bloodworm
Glycera sp.

**Atlantic
Jackknife Clam**
Ensis directus

coastal development has reduced suitable nesting areas. Gulls nest on offshore islands along the coast or on the rare isolated sandspits that still exist.

In summer, terns may be seen off most beaches along Outer Lands coasts, particularly when the adults and newly fledged young wander the shorelines and coastal waters in late summer in flocks before they migrate south. Terns look like small, delicate gulls with long, swallowlike tails (see illustrations, pp. 138–39). They hunt small fish such as Atlantic Silversides, Blueback and Atlantic Herring, and Sand Lance by diving onto schools of fish near the surface. Historically terns nested both on offshore islands and along beaches and sandspits of the Cape and Islands and Long Island coasts. As beaches were developed and became summer playgrounds, terns (except for Least Terns) were driven off most mainland nesting sites, and now the only large colonies of Common and Roseate Terns are on Falkner Island off Guilford, Connecticut, on Great Gull Island at the far eastern end of Long Island Sound, and on Monomoy Island south of Chatham on Cape Cod.

Ray Hennessy

Our endangered beach-nesting birds

Two beach-nesting birds are among North America's most endangered species, due to the almost total loss of their former nesting grounds. The Least Tern and the Piping Plover both nest in lightly vegetated upper beach and dune areas—precisely the types of areas that are mostly buried now beneath coastal houses or have been converted to recreational beaches. In the past decade both species have received more attention and protection of their nesting grounds in places such as Barnstable's Sandy Neck and at many of the ocean and bay beaches of Cape Cod National Seashore. On the south shore of Long Island there are protected beach nesting areas at Fire Island National Wildlife Refuge and Jones Beach State Park. *If you visit beaches with reserved nesting areas, please obey the posted signs* and stay away from the fenced-off areas, and never let a dog on the beach (with or without a leash) during the late spring and summer months when birds are nesting. This small accommodation to these beach-nesting birds has made a real difference in their nesting success rates.

A pair of Least Tern parents feed a Sand Lance (*Ammodytes americanus*) to their chick. Least Tern nests are minimal scrapes in the upper beach and are very easy to miss. **Please stay out of marked nesting areas on the beach.**

The rare and endangered Roseate Tern (*Sterna dougallii*) nests at scattered beach and island sites throughout the Outer Lands, usually in the company of much larger Common Tern colonies, such as those on Monomoy Island, off Chatham, and on Great Gull Island, off the tip of New York's Orient Point.

Forster's Tern
Sterna forsteri
15 in.

Common Tern
Sterna hirundo
14 in.

Least Tern
Sternula antillarum

9 in.

Black Tern
Chlidonias niger
(Nonbreeding plumage)

10 in.

Roseate Tern
Sterna dougallii
15 in.

Steve Byland

Piping Plovers (*Charadrius melodus*) are so well camouflaged that they are easy to miss even when you are looking for them. Their nests on the upper beach are even easier to miss, so ***please stay out of marked beach bird nesting areas*** to avoid inadvertently harassing these endangered birds or destroying their nests.

Caspian Tern
Hydroprogne caspia
21 in.

Black Skimmer
Rynchops niger
18 in.

Royal Tern
Thalasseus maximus
20 in.

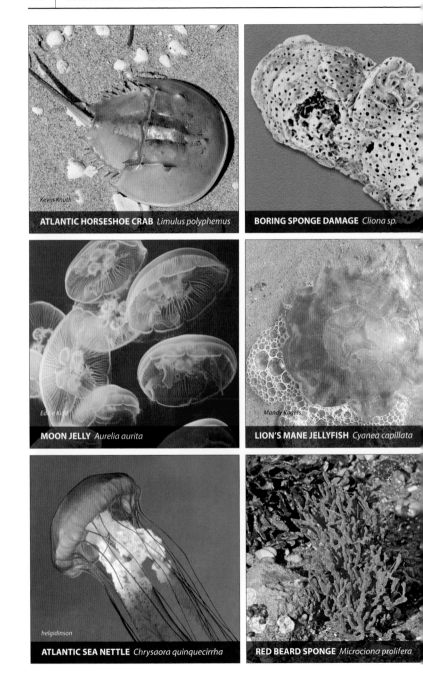

ATLANTIC HORSESHOE CRAB *Limulus polyphemus*

Kevin Knuth

BORING SPONGE DAMAGE *Cliona sp.*

MOON JELLY *Aurelia aurita*

Eddie Kidu

LION'S MANE JELLYFISH *Cyanea capillata*

Mandy Rogers

ATLANTIC SEA NETTLE *Chrysaora quinquecirrha*

helgidinson

RED BEARD SPONGE *Microciona prolifera*

PER SHELLS *Crepidula fornicata*

SAND DOLLAR *Echinarachnius parma*

hereswendy

EN FLEECE *Codium fragile*

SEA LETTUCE *Ulva lactuca*

KWEED *Fucus distichus*

KNOTTED WRACK *Ascophyllum nodosum*

NORTHERN MOON SNAIL
Euspira heros

RIBBED MUSSEL
Geukensia demissa

KNOBBED WHELK
Busycon carica

BAY SCALLOP
Argopecten irridians

BLUE MUSSEL
Mytilis edulis

ATLANTIC SURF CLAM
Spisula solidissima

CHANNELED WHELK
Busycon canaliculatus

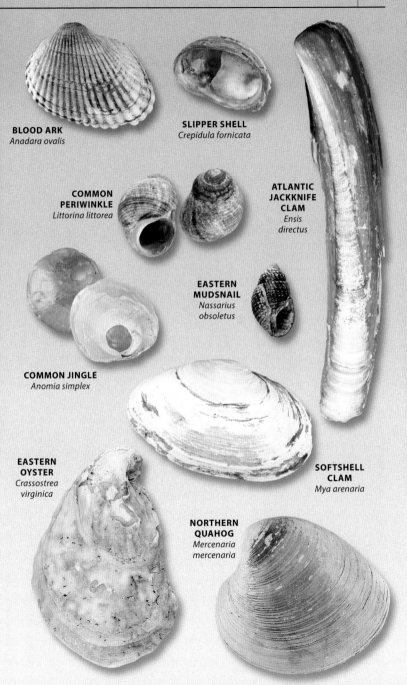

BLOOD ARK
Anadara ovalis

SLIPPER SHELL
Crepidula fornicata

**COMMON
PERIWINKLE**
Littorina littorea

**ATLANTIC
JACKKNIFE
CLAM**
*Ensis
directus*

**EASTERN
MUDSNAIL**
*Nassarius
obsoletus*

COMMON JINGLE
Anomia simplex

**EASTERN
OYSTER**
*Crassostrea
virginica*

**SOFTSHELL
CLAM**
Mya arenaria

**NORTHERN
QUAHOG**
*Mercenaria
mercenaria*

CANADA GOOSE
Branta canadensis

Common goose of coastal waters, inland ponds, bays, and marshes. Widely used as an ornamental goose for ponds and other freshwater habitats. These introduced birds are difficult to separate from the true wild and migratory Canada Geese. Introduced birds are more often than not the large subspecies called the Common Canada Goose (length typically 45 in., 111 cm), whereas wild Canada Geese, which migrate south in fall and return north to nest in spring, are usually a mix of smaller races (length typically 36 in., 91 cm). Found throughout United States and Canada. On East Coast, range extends to northern Florida and expands south each year. **Length:** 36–45 in. (91–111 cm). **Wingspan:** To 60 in. (1.5 m).

Common	J	F	M	A	M	J	J	A	S	O	N	D

BRANT
Branta bernicla

Small coastal and offshore goose resembling a small Canada Goose. However, back is darker brown to almost black, and black neck color extends down to a sharp cutoff mark on lower breast. Instead of the Canada Goose's bold chinstrap, the Brant has a small necklace of white lined with black. Bill is small and black. Breeds in the far north above Hudson Bay and migrates to coastal waters for fall and winter. Principal winter grounds range from the New England coast south to the northernmost coast of Georgia. There are scattered inland records and increased sightings farther south along the East Coast and on the Gulf Coast. **Length:** 26 in. (66 cm). **Wingspan:** 42 in. (107 cm).

Note: In the 1970s and 1980s the Brant population was in serious decline because its favorite food, Eelgrass (*Zostera*), was decimated by a fungus. Fortunately, the birds gradually altered their diets and now they feed on the marine algae Sea Lettuce (*Ulva lactuca*) and the fresh shoots of Saltwater Cordgrass (*Spartina alterniflora*), and the Brant population of the East Coast has begun to recover.

Spring & fall	J	F	M	A	M	J	J	A	S	O	N	D

BRANT

CANADA GOOSE

BRANT

Sexes are alike in all plumages

Sexes are alike in all plumages

GREAT BLACK-BACKED GULL   *Larus marinus*

A common bird of harbor areas and shorelines. Of all the large inshore gulls, this species will follow boats out to sea the farthest. The Great Black-Backed Gull is gradually extending its range down the Eastern Seaboard toward Florida.

Description: The largest gull of the shoreline. Jet-black back contrasts sharply with pure white underparts. White border shows from wingtips along rear edge of wings when in flight. Heavy yellow bill with red spot. Flesh-colored legs. First-year immatures can usually be told by massive size and by pale head and rump contrasting with brown back and underparts. **Length:** 30 in. (76 cm). **Wingspan:** 65 in. (1.7 m).

Habits: Aggressive and territorial with other birds, but often nests peacefully in mixed-species colonies with other gulls. Found throughout the northeast Atlantic Coast all year north of North Carolina and in winter south to northern Florida. Casual to the Gulf Coast.

Similar species: Superficially similar to the Herring Gull in plumage, but much larger, and note the much darker back, massive bill, and bulkier profile. The uncommon Lesser Black-Backed Gull also has an almost black back but is a bit smaller than a Herring Gull.

Common	J	F	M	A	M	J	J	A	S	O	N	D

Great Black-Backed Gull
First winter

Massive dark bill

Checkered effect of high-contrast plumage pattern

Herring Gull
First winter

Dull brown mantle lacks contrast

Smaller bill, light at the base

Length: 30 in.

Length: 25 in.

Comparison of young Herring and Great Black-Backed Gulls, often seen in mixed flocks on beaches

GREAT BLACK-BACKED GULL

Third winter

The stark black-and-white contrast of the adult is often the best field mark at a distance

Adult winter

The largest gull in North America

Second winter

First winter

Third winter

First winter

Heavy bill at all ages

Adult breeding

Pale pink legs at all ages

HERRING GULL

Larus argentatus

Description: An abundant bird—the gull most people think of when they think of a seagull. Large, with gray back and black wingtips with white spots. White head and underparts. Pink-tan legs. Yellow bill with blood-red spot on lower mandible. Yellow eye. In winter plumage, head is streaked brown with a dark eye line, giving the face a stern appearance. First-year immatures are chocolate brown with lighter speckles. Herring Gulls reach full adult plumage after four years. Plumage may be distinguished by year of age until adult plumage is reached. **Length:** 25 in. (64 cm). **Wingspan:** 58 in. (1.5 m).

Habits: An aggressive, opportunistic bird, readily adapting to both natural and man-made environments from well inland to miles from shore at sea. Will follow fishing boats well away from land. On the coast will often pick up shellfish and crabs and drop them from a height to crack their shells. Ranges along the entire East Coast in winter and from Maritime Canada to the Carolinas year-round.

Similar species: The similar-looking Ring-Billed Gull is smaller, with a more delicate bill. Also see the comparison of first-year birds, p. 150.

Abundant	J	F	M	A	M	J	J	A	S	O	N	D

Great Black-Backed Gull
Length: 30 in.

Herring Gull
Length: 25 in.

Ring-Billed Gull
Length: 18 in.

The larger white-headed gulls are superficially similar but separate distinctly by size. The Great Black-Backed Gull is a much more massive bird than the Ring-Billed Gull.

HERRING GULL

Dull brown mantle and lighter brown breast lacks contrast

First winter

Adult

Second winter

Gray central mantle and partially gray wings

Third winter

Similar to adult winter, but with darker head and tail

Second winter

Adult breeding

First winter

Pink legs at all ages

RING-BILLED GULL

Larus delawarensis

Description: A sleek, medium-sized gull of harbors, shorelines, and shopping center parking lots. Adults easily identified by distinct ring around bill. Very common in our area during the colder months. Gray back. Greenish yellow legs. Black wingtips spotted with white; black color extends along fore edge almost to wing bend. Reaches full adult plumage after three years. **Length:** 18 in. (46 cm). **Wingspan:** 48 in. (1.2 m).

Habits: A flexible, opportunistic species that has done very well in adapting to human development of the coastline. Mixes with other gulls in harbors and in large flocks resting on breakwaters and sandy shores. Will follow boats, hanging in the wind just astern and looking for handouts. Found throughout the Outer Lands. Ranges along the entire Atlantic and Gulf Coasts in the colder months. Breeds mostly in north-central Canada in summer.

Similar species: The Herring Gull is the most similar (see below and pp. 148–49). If you get a chance to see the two species side by side, note the much smaller, lighter body and more delicate features of the Ring-Billed Gull.

Abundant	J	F	M	A	M	J	J	A	S	O	N	D

Herring Gull
First winter
Length: 25 in.

Diffuse, dark bill tip

Mottled brown back

Ring-Billed Gull
First winter
Length: 18 in.

Much gray in the back

Well-defined black tip

Comparison of first-winter Ring-Billed and Herring Gulls

RING-BILLED GULL

A more contrasting pattern on mantle than first-winter Herring Gull

First winter

Adult breeding

Second winter

Gray mantle, showing minimal or no brown remnants

Tail lighter than the similar second-winter Herring Gull

Relatively small bill at all ages

First winter

Adult winter

Bill ring

Adult breeding

Yellow legs at all ages

LAUGHING GULL

Larus atricilla

Description: A trim gull, with black head and deep gray back. Blood-red bill and legs. No white in wingtips. Note broken white ring around eye. In winter plumage, hood fades to dark patch at back of head. Immature shows a black band at end of tail feathers. **Length:** 17 in. (43 cm). **Wingspan:** 40 in. (1 m).

Habits: This abundant gull's laughing call is a familiar sound from the mid-Atlantic Coast southward, and the Laughing Gull is becoming more common in the Northeast as the climate warms. Will follow inshore boats, hanging above the stern in search of handouts. Ranges along the entire East Coast in the warmer months and from Cape Hatteras south year-round.

Similar species: Bonaparte's Gull also has a black head in breeding plumage but is much smaller and more ternlike, and Bonaparte's Gull rarely mixes with Laughing Gulls.

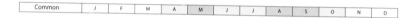

Common	J	F	M	A	M	J	J	A	S	O	N	D

Ring-Billed Gull
First winter
Length: 18 in.

Herring Gull
First winter
Length: 25 in.

Laughing Gull
First winter
Length: 17 in.

Comparison of immature Ring-Billed, Herring, and Laughing Gulls

LAUGHING GULL

First winter

Shows brown in wing mantle

Heavy terminal band

Adult breeding

Second winter

Very light terminal band

Black wingtips

First winter

Adult winter

Dark bill

Legs dark gray

Legs dark gray

Dark red bill

Adult breeding

Legs dark red in breeding plumage

BONAPARTE'S GULL

Larus philadelphia

Description: A small, almost ternlike gull. Very buoyant when sitting on the water. Black head in breeding plumage. Wings in all plumages show a distinct white wedge on outer edge. Blood-red bill, deep pink or red legs. In winter plumage, black hood is reduced to a black smudge behind eye. **Length:** 13 in. (33 cm). **Wingspan:** 33 in. (84 cm).

Habits: A fall-through-spring visitor to the Atlantic and Gulf Coasts. At times found well offshore, where it mixes with true oceanic birds. Bonaparte's is a small gull that doesn't often mix with the larger gull species, preferring to stand apart from mixed flocks on beaches and sandbars—a useful tip for spotting small flocks of these gulls. Ranges in winter throughout the Atlantic and Gulf Coasts and lower Mississippi River Valley.

Similar species: The small size and delicate bill separate it from the more common and much larger Laughing Gull. The small size, rounded, almost dovelike profile, and delicate beak are distinctive.

Spring & fall	J	F	M	A	M	J	J	A	S	O	N	D

Laughing Gull
Winter (nonbreeding)
Length: 17 in.

Bonaparte's Gull
Winter
Length: 13 in.

Comparison of winter Bonaparte's and Laughing Gulls

BONAPARTE'S GULL

First winter

Dark wingtips with light band

Dark wing bars

Tail band

White wedge near wingtips

Adult breeding

Adult winter

Cheek spot

Delicate bill

First winter

Adult breeding

Deep pink or red legs

ROSEATE TERN   *Sterna dougallii*

Pale gray to almost white. In breeding plumage, breast has a faint pink cast. Black bill with a deep red base. Long tail feathers form a deeply forked tail. In flight, wings are a clear, very pale gray. Feeds in inshore waters. An endangered species, common only around its few remaining breeding colonies. **Length:** 15 in. (38 cm). **Wingspan:** 29 in. (73 cm).

Rare & local	J	F	M	A	M	J	J	A	S	O	N	D

LEAST TERN   *Sternula antillarum*

The smallest tern. White forehead mark is present even in summer. The only local tern with a yellow bill, tipped with black. Immatures have especially noticeable dark leading edges on upper wings. An inshore bird; nests on beaches. Uncommon except near colonies. **Length:** 9 in. (23 cm). **Wingspan:** 20 in. (51 cm).

Uncommon	J	F	M	A	M	J	J	A	S	O	N	D

PIPING PLOVER   *Charadrius melodus*

Very pale. Plain face without dark cheeks. In breeding plumage, black bar across forehead and black collar around neck. These fade to pale buff in winter. Often sneaks away when approached, stopping and looking over its shoulder, relying on its sandy color for camouflage. Call "peep-low." **Length:** 7.5 in. (19 cm). **Wingspan:** 19 in. (48 cm).

Uncommon	J	F	M	A	M	J	J	A	S	O	N	D

COMMON TERN   *Sterna hirundo*

Gray back and upper wings, white belly and underwings. Black cap extends down nape. Orange-red bill with black tip. In flight, gray wedge at center of back and darker wingtips. In winter plumage, white forehead and front half of crown. Common near nesting colonies and in early fall migration. **Length:** 14 in. (36 cm). **Wingspan:** 30 in. (76 cm).

Common	J	F	M	A	M	J	J	A	S	O	N	D

AMERICAN OYSTERCATCHER   *Haematopus palliatus*

A large black-and-white shorebird with an unmistakable bright orange bill. No other shorebird in the area has a similar bill. It is flattened like a knife blade and inserted into bivalves such as oysters to pry open their shells. **Length:** 18 in. (46 cm). **Wingspan:** 32 in. (81 cm).

Uncommon	J	F	M	A	M	J	J	A	S	O	N	D

BLACK SKIMMER   *Rynchops niger*

A medium-sized relative of gulls and terns, with a unique, long lower mandible used to skim surface waters for small fish. Dark, almost black back; white underparts. Large, bright red bill with black tip. Most often noticed skimming across the surface of shallow coastal waters. Very rare nester near Cape and Islands. **Length:** 18 in. (46 cm). **Wingspan:** 44 in. (1.1 m).

Uncommon	J	F	M	A	M	J	J	A	S	O	N	D

Shown at same scale

Black bill with deep red at base in breeding season

ROSEATE TERN

Very white overall

Long tail projects well beyond folded wings

Slimmer body than the very similar Common Tern

LEAST TERN

PIPING PLOVER

Tiny; only local tern with yellow bill

COMMON TERN

Bright red-orange bill

AMERICAN OYSTERCATCHER

BLACK SKIMMER

GREAT BLUE HERON   *Ardea herodias*

Our largest heron is common in both freshwater and saltwater marshes and visits beaches and tidal flats throughout the area. Immatures have streaking on throat and upper chest, lack long head plumes, and are a more uniform gray all over. Common in fresh and saltwater wetlands and tidal flats. **Length:** 46 in. (1.2 m). **Wingspan:** 72 in. (1.8 m).

Common	J	F	M	A	M	J	J	A	S	O	N	D

LESSER YELLOWLEGS   *Tringa flavipes*

Lesser Yellowlegs often occur in mixed flocks with Greater Yellowlegs, making it much easier to tell them apart. The Lesser Yellowlegs is more delicate-looking than the Greater Yellowlegs. When seen alone, separation is difficult. Call: a single "tsip" or double "tu-tu." Common on beaches, on tidal flats, and in salt marshes throughout the Northeast coast. **Length:** 10.5 in. (25 cm). **Wingspan:** 25 in. (64 cm).

Spring & fall	J	F	M	A	M	J	J	A	S	O	N	D

GREATER YELLOWLEGS   *Tringa melanoleuca*

Larger and less delicately proportioned than Lesser Yellowlegs. Upper part brown flecked with white. Underparts show heavier barring on flanks than Lesser in breeding plumage. Often feeds running about with its bill in water swinging from side to side. Call: a loud "teu teu teu." Found across United States in migration. Winters in the Southeast from Virginia south to Florida and Gulf region. **Length:** 14 in. (36 cm). **Wingspan:** 28 in. (71 cm).

Spring & fall	J	F	M	A	M	J	J	A	S	O	N	D

WILLET   *Tringa semipalmata*

In flight the large white wing stripes are unmistakable. Grayish brown with distinct back and underpart barring in breeding plumage. Very vocal, calling its name repeatedly: "pill-will-willet!" Favors protected shoreline and upper beach areas for feeding and nesting and is also found in salt marshes. **Length:** 15 in. (38 cm). **Wingspan:** 26 in. (66 cm).

Uncommon	J	F	M	A	M	J	J	A	S	O	N	D

GREAT BLUE HERON
in flight

Shown at same scale

GREAT BLUE HERON

LESSER YELLOWLEGS

Breeding

Delicate bill, about as long as width of head

Fall and winter

More substantial bill, longer than width of head

Fall and winter

GREATER YELLOWLEGS

Breeding

Breeding plumage

WILLET

Large white wing stripes

Fall and winter

SANDERLING — *Calidris alba*

Anyone who spends time at the beach is familiar with these small sandpipers chasing the waves in and out like tiny wind-up toys. In winter plumage, very white-looking with gray back and jet-black legs. In breeding plumage, rich cinnamon red around head and neck. **Length:** 8 in. (20 cm). **Wingspan:** 17 in. (43 cm).

Spring & fall	J	F	M	A	M	J	J	A	S	O	N	D

SEMIPALMATED PLOVER — *Charadrius semipalmatus*

Black collar around neck, black mask through eye, and white forehead. Bill yellow with black tip. Legs yellowish orange. In flight shows black-tipped tail with white sides. Call: a sharp "chu-wee." **Length:** 7.5 in. (19 cm). **Wingspan:** 19 in. (48 cm).

Spring & fall	J	F	M	A	M	J	J	A	S	O	N	D

DUNLIN — *Calidris alpina*

Rusty red back, black underbelly. Long bill curves down at tip. In nonbreeding plumage changes to nondescript brown-gray plumage on back, head, and chest with paler underparts. **Length:** 9 in. (23 cm). **Wingspan:** 17 in. (43 cm).

Spring & fall	J	F	M	A	M	J	J	A	S	O	N	D

LEAST SANDPIPER — *Calidris minutilla*

A small sandpiper that prefers muddy areas toward the rear of the beach and pool edges with grass rather than the sandy beach itself. Crown and back are brownish rather than grayish. Adopts a crouched posture as it feeds. **Length:** 6 in. (15 cm). **Wingspan:** 13 in. (33 cm).

Spring & fall	J	F	M	A	M	J	J	A	S	O	N	D

BLACK-BELLIED PLOVER — *Pluvialis squatarola*

Breeding plumage boldly marked black and white. Underparts mainly black, with mottled light gray back that runs up neck to white cap. Short, black, heavy bill. Nonbreeding plumage predominantly gray, but may show areas of black on underparts. Call a distinctive mournful "pee-ooo-wee." **Length:** 11.5 in. (29 cm). **Wingspan:** 29 in. (74 cm).

Spring & fall	J	F	M	A	M	J	J	A	S	O	N	D

KILLDEER — *Charadrius vociferus*

Two black neck bands, full white collar, and white at bill base and over eye. Long, tapered body, brown on back and white below. Common in most open environments, beaches, grassy fields, and fields near marshes. **Length:** 10.5 in. (27 cm). **Wingspan:** 24 in. (61 cm).

Uncommon	J	F	M	A	M	J	J	A	S	O	N	D

SEMIPALMATED SANDPIPER — *Calidris pusilla*

Relatively short, dark bill and dark legs. We tend to see them mostly in migration, where the plumage is often a dull, nondescript gray. Spring birds in breeding plumage show much more rusty red tones on back and breast and have more contrast in dark patterning of back. **Length:** 6.5 in. (17 cm). **Wingspan:** 14 in. (36 cm).

Late summer & fall	J	F	M	A	M	J	J	A	S	O	N	D

Pale gray in non-breeding plumage

Shown at same scale

SANDERLING

SEMIPALMATED PLOVER

Bill droops noticeably at tip

White belly with brown breast (nonbreeding plumage)

DUNLIN

LEAST SANDPIPER

Generally rufous back and flank

Breeding

Fall and winter

BLACK-BELLIED PLOVER

KILLDEER

SEMIPALMATED SANDPIPER

Short, dark bill, dark legs, dull gray in fall and winter (nonbreeding plumage)

When you approach a Killdeer nest, the birds will often do a broken wing display, pretending to be injured to draw you away from the nest.

The main hiking trail at Barnstable's Sandy Neck, the best place on the Cape to explore both dune environments and the huge salt marshes adjacent to Sandy Neck.

SAND DUNES

Today's dunes are an ecological community that hinges on a single plant, American Beachgrass, without which most dune sand would have long since blown back into the sea.

Beaches and dunes are made of sand, and along the New England coast both environments are inextricably linked. Outer Lands dunes originate from glacial till, cut from the land by weather and water, and mixed and sorted by powerful ocean waves, which then throw the most durable small mineral grains back on the shore as beach sand. Onshore winds constantly move sand up the beach, where the grains are caught by the stems and leaves of plants. Here they form the primary dunes that give rise to both sandspits and larger dune communities. No matter how large, these dune systems were thus all born on beaches. Beaches are sculpted by moving water and wind, but dunes are the province of the wind.

So how do you tell a dune environment from the usual range of plants that inhabit the upper beach? In large dune areas such as the Provincelands or Sandy Neck, the answer is obvious, with sand hills as far as the eye can see. But most dune areas are far smaller, and the difference between the upper beach and dune communities is more subtle.

First, there's the physical distance from the ocean waves and the influence of salt spray. Upper beach communities share many of the same hardy plants as true dune environments, but beaches are defined by the short distance to water and waves, where plants must be able to survive and thrive with a constant mist of salt water on their leaves, as well as an occasional salty drenching of their roots by perigean high tides

A Pitch Pine (*Pinus rigida*) forest and dune community just west of the entrance booth to the Sandy Point Beach Park parking lot in Barnstable. Note the extensive ground cover on either side of the path. Mature dune community like this have a ground cover of Gray Reindeer Lichen (*Cladonia rangiferina*), Beach Heather (*Hudsonia tomentosa*), and other assorted mosses, lichens, and grasses.

(king tides). A true northeastern dune community is defined by the presence of plants that can survive the heat, dryness, and sandblasting of living on sand but have protection from most salt spray by distance and a line of primary dunes.

With distance from salt water a complex and distinctive community of lichens, fungi, grasses, shrubs, wildflowers, and trees can develop and thrive. Watch the ground as you walk (preferably on an established footpath) from the beach and over the first high mounds of sand atop the beach, where bare sand covers the ground between plants. In the Outer Lands you know you are in a true mature dune community through the presence of two key ground covers: Gray Reindeer Lichen and Beach Heather. The common upper beach herbs and grasses, Northern Bayberry bushes, Beach Plums, and other upper beach plants also thrive on dunes, but only when you see a mixed carpet of Gray Reindeer Lichen, other lichens, small patches of moss, and Beach Heather in the open areas have you arrived in a true dune environment.

The origin and geography of dunes

Dunes by the sea are iconic images of the Outer Lands, but extensive dune environments are not common. Three of the four largest Outer Lands dune areas are on Cape Cod: the Provincelands at the northern tip of Cape Cod, Monomoy Island just south of Chatham, and Sandy Neck peninsula in Barnstable. Long Island's Fire Island also has a substantial dune community. Other dune habitats are scattered and much smaller and occur mostly where sandspits and barrier islands

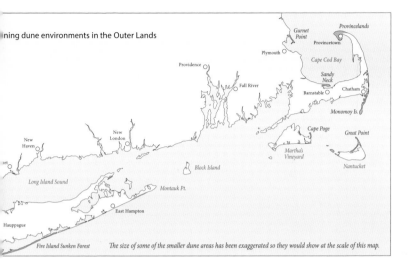

...ning dune environments in the Outer Lands

The size of some of the smaller dune areas has been exaggerated so they would show at the scale of this map.

are large enough to provide sandy areas with little exposure to salt spray. There is irony here: these signature coastal wildlands are both natural ecologies and wholly artificial human works. Before the Pilgrims' arrival, dune communities were small and scattered, and the landscape was dominated by mature coastal hardwood forests, even in most sandy areas.

As with almost every other square foot of coastal habitat in New England and the islands, the major dune areas of the Provincelands and Sandy Neck have been heavily modified by human activities, mostly through clear-cutting and livestock grazing. Today it is difficult to imagine those vanished primeval forests. Wood End at the tip of Cape Cod was named for dense stands of woodland on Race Point. Today Wood End Light stands with sea on three sides amid another sea of sand. The Beech Forest just north of Provincetown hints at what these precontact sand plain forests were like, but even it is only about 130 years old, and no virgin forests remain on Cape Cod.

Although most inland forest areas have recovered from multiple clear-cuts over the past 400 years, forests on sand are particularly fragile environments—easy to destroy and very slow to recover. The earliest Europeans encountered mature hardwood coastal forests over most of the Outer Lands, including the Provincelands and Sandy Neck, with smaller mature sand plain and pine barrens communities dominated by Pitch Pines. As settlers logged off the virgin forests and then allowed their sheep and cattle to graze over the newly

Extensive dune areas are not common in the Outer Lands. Most dune environments are small interior sections of large beaches, sandspits, and barrier islands.

Wood End Light at the very tip of Cape Cod. Wood End was named in early colonial times for the extensive forests of the Race Point area, a thing scarcely imaginable in today's dune environment.

A Pitch Pine dune forest in the Provincelands at the northern tip of Cape Cod.

A recently burned group of Pitch Pines near the Pilgrim Heights area of Cape Cod National Seashore in Truro. Although most of these trees were badly scorched, the area will quickly regenerate as the pines sprout buds both from under the bark of their trunks and from below ground level at the base of their roots.

opened fields, the woodland soils were exposed to the sun and wind and gradually blew away. Stripped of the stabilizing cover of vegetation, massive amounts of sand began to move. As early as 1714, residents of Provincetown grew alarmed at the extent of the newly barren sand areas in the Provincelands, and especially during winter storms, the mountains of sand moved across roads, buried outlying houses, and even threatened to inundate the town. State laws passed in 1725 forbade grazing in the dune areas and encouraged the planting of American Beachgrass and Pitch Pines. But not until the mid- to late 1800s were serious efforts made to restore the Provincelands with extensive plantings of Pitch Pines and American Beachgrass, and most of the Provincelands Pitch Pine forests originated less than 150 yeas ago.

The removal of the original hardwood forests and their replacement with Pitch Pines caused a further misery to nineteenth- and early twentieth-century Cape Cod residents: wildfires. These Pitch Pine forests were (and still are) susceptible to fires started accidentally by people or by lightning, and Cape Cod is one of the few areas in the Northeast where forest fires are a danger. However, with today's communication systems and much more sophisticated fire control techniques, major wildfires on the Cape are rare.

Today's dune areas have much more bare sand then they did in precontact times, and human traffic through dune areas still damages the fragile vegetation cover. When you visit dune communities, please obey any signs that restrict foot traffic through the dunes. A careless footstep could easily destroy decades' worth of lichen growth, exposing yet more bare sand to the wind.

Formation of dunes

Dunes occur naturally along sandy beaches. The upper beach above the wrack line receives significant salt spray, and this constant salt coating kills all but a sparse sprinkling of grasses and other plants. However, just up the beach the salt spray lessens, plants grow more densely, and the stems and leaves of American Beachgrass, Beach Clotbur, Searocket, and other beach plants begin to trap windblown sand. As the sand piles up around the plants, they respond by growing upward, and soon a foredune forms at the top of the beach. Foredunes don't attract plants: the plants themselves create the foredune and give the dune stability through their roots and rhizomes under the surface. On the upper beach look for sand shadows next to beach plants (see illustration, p. 170). As the wind-blown sand meets a plant on the upper beach, the wind slows, dropping grains of sand on the leeward side of the plant. Soon a stretched oval shadow of piled sand forms next to the plant. In this way plants build the dune environment by collecting sand around them and growing upward to stay atop the pile.

Wind and sand

Several processes transport sand grains. Strong winds pick up surface grains and transport them in short leaps across the surface in an action called saltation (from the Latin for "jumping"). Wind also can shove the grains of sand directly along the beach surface in a process called surface creep.

Wind is the dominant factor in dune environments. With little physical shelter, the dune plant community is exposed to strong northeasterly winds in winter and hot, drying south-westerly winds in summer. Wind constantly shifts the surface sand grains, which are typically caught and held in areas of

A panorama of the Provincelands dunes and Pitch Pine forests from the Province Lands Visitor Center at Cape Cod National Seashore.

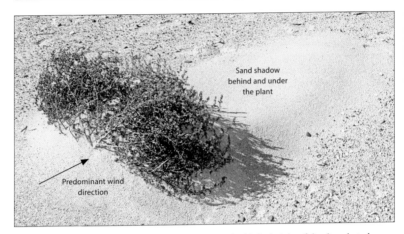

Sand shadow
behind and under
the plant

Predominant wind
direction

A sand shadow cast by a Common Saltwort (*Salsola kali*) on an upper beach. Even a single small plant can trap significant quantities of sand.

vegetation, helping to build the height of the dune but also exposing dune plants to low-grade sandblasting every day.

Even in a relatively moist climate like the Outer Lands, dunes are a dry environment because of how quickly surface water drains away. What rain the dunes receive quickly drains in the porous sand toward a groundwater table that may be six to ten feet or more below the surface. Salt spray can damage plants that lack tough leaves, but the constant mist of salt spray also brings minerals to an environment where most nutrients are washed down into the fast-draining sand. Nitrogen especially is in limited supply in dune areas, contributing to poor soil productivity.

Dune environments receive intense light and heat under the summer sun, increasing the rate of evaporation. The dry, light-colored sand of bare dunes can act as a parabolic reflector in dune swales, concentrating the sun's rays and bringing

The layered beds of a sand dune at Sandy Point in Barnstable.

heat levels well above that of the open beaches nearby. On more mature and stable dune fields farther back from the shoreline, larger plants such as Northern Bayberry bushes, Black Cherries, and Eastern Redcedar Junipers provide shade and act as windbreaks. In the Cape and Islands and on eastern Long Island you may see small areas of the classic dune ground cover communities of mixed Bearberry, Broom Crowberry, Bear Oak, Gray Reindeer Lichen, and Beach Heather.

The brilliant white appearance of dune sand is no accident: dune sand is lighter, finer grained, and whiter than typical beach sand in the Outer Lands. As the wind blows sand grains inland it sorts out mineral particles, and smaller grains of extremely durable but relatively lightweight quartz and feldspar move inland the farthest.

Dune plant communities

Dune plants must have tough, flexible leaves, able to withstand the mechanical strain of whipping in the wind, along with impervious waxy or hairy leaves that help to limit moisture loss. Many dune plants, such as American Beachgrass, also have leaves that curl in high heat, limiting moisture loss.

The two dominant dune forms in the Outer Lands are conventional coastal dunes (top figure) and the much larger parabolic dunes (bottom figure) that form where large areas of free sand surface are exposed to winds from a consistent direction. Most dune areas on the Cape and Islands are conventional coastal dunes. The huge sand dunes just north and northwest of Pilgrim Lake in the Provincelands are parabolic dunes, formed primarily by the dominant northwest winds of winter on the Outer Cape.

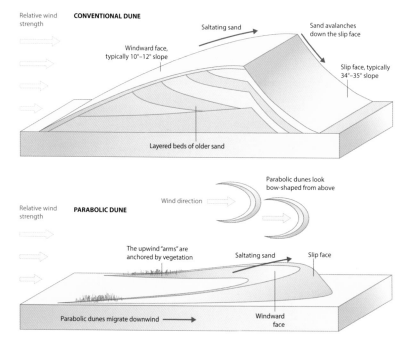

CONVENTIONAL DUNE

Relative wind strength

Saltating sand

Sand avalanches down the slip face

Windward face, typically 10°–12° slope

Slip face, typically 34°–35° slope

Layered beds of older sand

PARABOLIC DUNE

Relative wind strength

Wind direction

Parabolic dunes look bow-shaped from above

The upwind "arms" are anchored by vegetation

Saltating sand

Slip face

Parabolic dunes migrate downwind ⟶

Windward face

Rugosa Rose (*Rosa rugosa*) is often the first shrub to appear in foredune plant communities, where its toughness and resistance to salt spray helps stabilize foredunes throughout the Northeast.

A classic foredune community, dominated by Seaside Goldenrod (*Solidago sempervirens*) and American Beachgrass (*Ammophila breviligulata*).

Many perennial dune flowering plants and grasses spread aggressively through underground stems called rhizomes. This both increases the area a plant can cover and allows the plant to respond quickly to burial by drifting sand through rapid upward growth from the rhizomes.

Like most beach plants, many dune plants are annuals. Annual plants invest all of their reproductive energy in seed production, and they are most common in constantly changing or disturbed environments where their seeds can rapidly germinate to take advantage of favorable conditions. With no permanent stems or roots to protect during the harsh winter months, annual plants spread their seeds widely, placing their evolutionary bets on next year's seedlings. Given that such annuals as Seaside Spurge, Common Saltwort, and Searocket are so common on beaches and dunes, the strategy has been successful.

Foredune communities

Foredunes are simply a higher continuation of the upper dry beach and are usually the first place that significant amounts of plants appear. Most foredune plants are the same ones you'll find scattered on the upper beach—American Beachgrass, Seaside Spurge, Common Saltwort, Seabeach Orach, Searocket, and Seaside Goldenrod—but here they are more numerous. Other dune community plants begin to appear on the foredune, including Dusty Miller, Beach Pea, and Umbrella Sedge, a grasslike plant. The introduced Rugosa Rose is the hardiest and most salt-resistant of the dune shrubs

and is common on foredunes and sandy areas throughout the New England shoreline. Many common inland plants and wildflowers are also hardy enough to live on foredunes, well back from the beach: Yarrow, Evening Primrose, Switchgrass, Virginia Creeper, and Poison Ivy.

Less exposure to salt spray is the main thing that distinguishes true dune environments from the saltier foredune communities. All the foredune plants mentioned above are also present on dunes, but the primary thing you'll notice about true dune environments is the presence of shrubs and stunted trees. Rugosa Rose, Shining (Winged) Sumac, Northern Bayberry, Beach Plum, and Black Cherry are all common dune shrubs. Black Cherry is also a common forest-edge tree that can grow quite tall, but on dunes the salt spray and lack of water keep the Black Cherries small and shrublike. Quaking Aspen appears on dunes that are well away from salt spray, and if present the aspens are often the tallest trees in the dunes. Aspens are more common where dune fields transition to bordering maritime forests, and Black Locust trees will tolerate the very dry conditions of beaches. In scattered locations you'll find American Holly growing wild in the dunes, although wild holly is more common on Long Island than on the Cape or more northern locations. New England's only cactus species also appears in dunes. Eastern Prickly-Pear is intolerant of salt spray or salt water on its roots and is usually found well back from the beach in areas sheltered from salt spray by shrubs or on rocky headlands near the coast. Along with the ubiquitous American Beachgrass, you'll see additional grasses

A grouping of the dominant dune shrubs: Eastern Redcedar Juniper (*Juniperus virginiana*), Beach Plum (*Prunus maritima*), and the nearly ubiquitous Northern Bayberry (*Myrica pensylvanica*). In wetter areas near beaches and salt marshes, Marsh Elder (*Iva frutescens*) often joins the mix.

Eastern Redcedar Juniper

Beach Plum

Northern Bayberry

Marsh Elder

AMERICAN BEACHGRASS *Ammophila breviligulata*

SEASIDE SPURGE *Chamaesyce polygonifolia*

COMMON SALTWORT *Salsola kali*

SEABEACH ORACH *Atriplex pentandra*

SEAROCKET *Cakile edentula*

SEASIDE GOLDENROD *Solidago sempervire*

STY MILLER *Artemisia stelleriana*

BEACH PEA *Lathyrus japonicus*

NDBUR *Cenchrus longispinus*

RUGOSA ROSE *Rosa rugosa*

CH HEATHER *Hudsonia tomentosa*

BEACH HEATHER, detail

SWITCHGRASS *Panicum virgatum*

VIRGINIA CREEPER *Parthenocissus quinquefol*

POISON IVY *Toxicodendron radicans*

CATBRIER *Smilax glauca*

EASTERN PRICKLY-PEAR *Opuntia humifusa*

YELLOW WILD INDIGO *Baptisia tinctoria*

Y REINDEER LICHEN *Cladonia rangiferina*

BRITISH SOLDIERS LICHEN *Cladonia cristatella*

HY BEARD LICHEN *Usnea strigosa*

HYGROSCOPIC EARTHSTAR *Astraeus hygrometricus*

RBERRY *Arctostaphylos uva-ursi*

BROOM CROWBERRY *Corema conradii*

SWEET FERN *Comptonia peregrina*

MARSH ELDER *Iva frutescens*

UMBRELLA SEDGE *Cyperus strigosus*

RED GOOSEFOOT *Chenopodium rubrum*

BEACH PLUM *Prunus maritima*

BEAR OAK *Quercus ilicifolia*

K CHERRY *Prunus serotina*

NORTHERN BAYBERRY *Myrica pensylvanica*

MON JUNIPER *Juniperus communis*

PITCH PINE *Pinus rigida*

E PINE *Pinus strobus*

SASSAFRAS *Sassafras albidum*

American Holly
Ilex opaca

Eastern Redcedar Juniper
Juniperus virginiana

and sedges here, including Switchgrass, Cheatgrass, Nutsedge, and Umbrella Sedge.

In protected dune areas well back from the beach, ground cover plants such as Bearberry, Broom Crowberry, and Beach Heather may form small heathlike carpets, especially when sheltered from wind and salt spray by such small trees and shrubs as Northern Bayberry, Black Cherry, Bear Oak, and Beach Plum. These dune heath areas are also home to a few distinctive lichens: Gray Reindeer Lichen and British Soldiers Lichen thrive on sandy soils. Also sometimes seen on bare patches of sand, especially after a rainstorm, is the Hygroscopic Earthstar. Two conifers are common: the Eastern Redcedar Juniper is the most common evergreen in many upper beach and salt marsh environments because of their tolerance of salt spray, but farther from the beach Pitch Pines become the dominant conifer. At the inland edges of dune areas, where sandy glacial soil replaces pure sand, White Pines will mix with Pitch Pines. As New England's tallest tree, White Pines often shade and outcompete Pitch Pines, but their shallower root systems make White Pines more vulnerable to storm winds, and it is often the hardy Pitch Pine that wins out in the competition with its loftier relative.

An appreciation of the Pitch Pine

If any species can be said to define the Outer Lands dunes, it is the Pitch Pine. Useless for sawmill lumber, a poor, smoky choice for firewood, and highly susceptible to forest fires, the humble Pitch Pine is also a wildly beautiful tree of almost unbelievable durability. What experience is more characteristic of the Outer Lands than to stand beside the ocean in a grove of these gnarled pines, hearing the sea wind hiss through their needles, with the sharp scent of pitch in the air?

The Pitch Pine is native to the sandy and rocky parts of the New England coast but is now much more common than it was in precontact times due to extensive plantings to stabilize sandy coastal areas left barren after centuries of human abuse. With its tolerance of extremely dry conditions and deep root systems that allow it to tap hidden supplies of groundwater, the Pitch Pine has been a near-perfect choice to restore dune environments. Those deep roots also stabilize Pitch Pines in storm winds and allow them to resist the tons of pressure from dune sand as the position of dunes shift with the wind.

Though highly vulnerable to wildfires, Pitch Pines are well adapted to fire-prone environments. They are one of the few conifers that can sprout from burned-over stumps. Even if a tree has completely burned to the ground, the surviving roots will send up new sprouts within weeks. Although Pitch Pines

A Pitch Pine in the dunes near the Beech Forest, just north of Provincetown.

Common Dune Shrubs

Marsh Elder
Iva frutescens

Groundsel Tree
Baccharis halimifolia

Northern Bayberry
Myrica pensylvanica

Beach Plum
Prunus maritima

Black Cherry
Prunus serotina

will tolerate some salt spray, trees near the beach or the edges of bluffs over the ocean will grow low across the ground or be stunted and salt-burned, yet still survive.

Dune swales

Dune swales are low areas in dunes that are more sheltered from the wind and have moister soils. Most dune swales are seasonal and hold standing water or wet soils only in cooler months. In summer the swales can be sometimes recognized by thin patches of Mud Rush (*Juncus pelocarpus*) and Canada Rush (*J. canadensis*) growing in low areas between dunes. Few other plants can survive a wetland that becomes a blazing-hot desert for three to four months of the year.

More reliably moist swale areas may contain a variety of the usual dune shrubs and small trees, as well as such salt marsh plants as Glasswort, Sea Lavender, Spike Grass, and Saltmeadow Cordgrass (see the next chapter, "Salt Marshes"). Besides the dune shrubs pictured at left you may see such common salt marsh border shrubs and grasses as Marsh Elder, Groundsel Tree, and Switchgrass. On the Cape and Islands small patches of wild American Cranberrybush and Highbush Blueberry sometimes grow in the wetter swales.

In many areas of the New England coast, small dune fields are often next to salt marshes on the landward side of sandspits, so salt marsh and dune environments often intermix. You can see this mixing of marsh and dune environments on the landward coast of Sandy Point in Barnstable, on the sandspit south of Coast Guard Beach in Eastham, on the pond side of Poge Point on Chappaquiddick Island, and on many other sandspits throughout the Outer Lands where a sandspit protects an adjacent salt marsh from ocean waves.

Sunken forests

A sunken forest is a forest within a dune swale or protected low valley on a barrier island, as in the most famous instance, the Sunken Forest at Fire Island National Seashore on the south coast of Long Island. Sunken forests occur in unusual circumstances in which a deep dune swale is within 400–600 yards of an ocean beach. The sunken, protected swale allows salt-averse trees to develop much closer to the beach than normal, but once the trees reach the height of the dune on the seaward side, they are cropped off by salt spray. The result is a miniature forest of dwarfed hardwood trees, usually 20–25 feet high, with tops that look as if a giant hedge trimmer cut all their foliage to the same height.

The Sunken Forest of Fire Island is dominated by three hardwood species: American Holly, Sassafras, and Shadbush (Canadian Serviceberry, *Amelanchier canadensis*). Small sunken

SH ELDER *Iva frutescens*

GROUNDSEL TREE *Baccharis halimifolia*

BUSH BLUEBERRY *Vaccinium corymbosum*

AMERICAN CRANBERRYBUSH *Viburnum trilobum*

LAVENDER *Limonium carolinianum*

GLASSWORT *Salicornia sp.*

The huge dunes of Sandy Neck are gorgeous but highly vulnerable to blow-outs in winter storms.

forests also occur in the Provincelands of Cape Cod, but wild American Hollies are much more unusual, except in areas near homes and buildings where they are often planted as decorative specimens. At the margins of a sunken forest you'll find Eastern Redcedar Junipers, Pitch Pines, and the usual range of dune shrubs and understory plants, with Northern Bayberry, Black Cherry, and Beach Plum as the most common species. Below the sunken forest trees you'll see Catbrier, Virginia Creeper, and the ubiquitous Poison Ivy, and often the three common sumac species: Staghorn, Smooth, and Shining (Winged) Sumac.

Animals on the dunes

Most beach animals and birds also frequent dunes, particularly if they nest on beaches. Our two most endangered beach nesters, the Piping Plover and the Least Tern, both sometimes use adjacent dune areas as nesting sites, particularly if the vegetation remains low and grassy and is not too thick. Other birds that nest in the transition areas from beaches to dunes include the Black Skimmer, American Oystercatcher, and Willet. Many songbirds also nest in dune areas, particularly the Northern Mockingbird, Common Grackle, American Robin, Song Sparrow, and in marshy swales the Red-Winged Blackbird.

In the shadows of dune plants small animals prowl, including large wolf spiders, Seaside Grasshoppers, field crickets, and (unfortunately) several species of ticks that mostly parasitize wild mammals, including White-Footed Deer Mice, Meadow Voles, Raccoons, and Red Foxes. Always wear long pants if you explore away from trails in sandspits and grassy or brushy

The central platform on the boardwalk through the Sunken Forest at Fire Island National Seashore.

ERICAN HOLLY *Prunus serotina*

CANADIAN SERVICEBERRY *Amelanchier canadensis*

SAFRAS *Juniperus virginiana*

STAGHORN SUMAC *Rhus typhina*

OOTH SUMAC *Rhus glabra*

SHINING (WINGED) SUMAC *Rhus copallinum*

Lone Star Tick
Amblyomma americanum

Black-Legged Tick
(often called the deer tick)
Ixodes scapularis

dune areas, and apply a DEET-based insect repellent on your clothes, socks, and shoes. Dune and beach vegetation often harbors the main carrier of Lyme disease, the Black-Legged Tick (often called the deer tick). The much more common Lone Star Tick can also carry diseases, so always check your clothing and any exposed skin when you exit wild shoreline environments.

The beaches and dunes of the New England coast are some of the best places to observe migrating insects in the fall months. On a clear September or October day with a brisk northwest wind sweeping them down to the coastline, hundreds of migrating Monarch butterflies, Green Darner dragonflies, and Black Saddlebags dragonflies move along the shores, often using the dune vegetation for rest and shelter. Unfortunately, these gorgeous flocks of migrating Monarchs may soon be a thing of the past. Over the past decade scientists have seen a major decline in the number of migrating Monarchs, which overwinter in just a few valleys in northern Mexico. Those Mexican valleys are now fairly well protected as conservation areas, and researchers are looking at the increased use of glyphosate herbicides by farmers since 2003. Monarch caterpillars feed on milkweeds, and the major reduction in milkweeds (particularly in midwestern farming areas) is suspected to be behind the 59 percent drop in overwintering Monarch populations in Mexico in 2012. So appreciate these black-and-orange beauties while we have them, because they will likely be less common in the future.

FALL COASTAL MIGRATORY INSECTS

Black Saddlebags
Tramea lacerata

Green Darner
Anax junius

Monarch
Danaus plexippus

ERICAN COPPER *Lycaena phlaeas*

CLOUDED SULPHUR *Colias philodice*

SIDE DRAGONLET *Erythrodiplax berenice*

FOWLER'S TOAD *Bufo woodhousei*

URNING CLOAK *Nymphalis antiopa*

VIRGINIA OPOSSUM *Didelphis virginiana*

PIPING PLOVER *Charadrius melodus*

LEAST TERN *Sternula antillarum*

BLACK SKIMMER *Rynchops niger*

WILLET *Tringa semipalmata*

SONG SPARROW *Melospiza melodia*

NORTHERN MOCKINGBIRD *Mimus polyglottos*

LOW-RUMPED WARBLER *Setophaga coronata*

AMERICAN ROBIN *Turdus migratorius*

RTHERN CARDINAL *Cardinalis cardinalis*

AMERICAN GOLDFINCH *Carduelis tristis*

FOX *Vulpes vulpes*

MEADOW VOLE *Microtus pennsylvanicus*

Raccoon
Procyon lotor

Eastern Cottontail
Sylvilagus floridanus

Striped Skunk
Mephitis mephitis

As in other environments, most animal activity in the dunes is at night, where the Raccoon, Virginia Opossum, Striped Skunk, Meadow Vole, and Eastern Cottontail are all common. Two less common dune specialists are Fowler's Toad and the Eastern Hognose Snake. Fowler's Toad is the only amphibian typically found on beaches and dunes, where it buries itself in sand under shrubs to escape the heat of the day. Hognose snakes are nonpoisonous and harmless but will sometimes perform an elaborate rearing and hissing display if surprised. Hognose snakes are very reluctant to bite humans even if handled (but please don't handle them) and are more likely to roll over and play dead if you approach them.

Dunes in winter

In winter dunes can be very productive and interesting areas for birding. Many northern species fly south to winter along the Atlantic Coast, and for species used to the wide expanses of Arctic tundra, beaches must seem like familiar territory. Snow Buntings, Lapland Longspurs, and Horned Larks are all small songbirds that nest in the high Arctic and winter on open fields, beaches, dunes, and marshes near coastlines.

Almost every winter a few spectacular Snowy Owls visit the marshes and beaches of the Outer Lands, but usually not in large numbers. In the winter of 2013–14 and again in the winter of 2014–15, however, many dozens of Snowy Owls were spotted throughout New England and the Atlantic Coast area, probably due to cyclical increases in the population of their lemming prey in the Arctic. In the breeding season of 2013 and apparently again in 2014, so many young Snowy Owls survived to fledge that many juvenile and young adult birds ventured far south of their normal wintering grounds in search of reduced competition from their peers and adult owls for food. Snowy Owls gravitate to open coastal beach and marsh areas that are similar to their normal tundra habitat. Snowys are solitary creatures in winter, and they can be difficult to spot against clumps of snow and ice. On winter beaches, dunes, and salt marshes these large, white owls will often take an exposed perch on a driftwood snag, post, or rock, looking for small mammals to capture and eat. Unlike their nocturnal cousin the Great Horned Owl, Snowy Owls are mostly daytime hunters that locate their prey by sight.

Snowy Owl
Bubo scandiacus

Dune forest with Pitch Pines, Sandy Neck Beach Park, Barnstable.

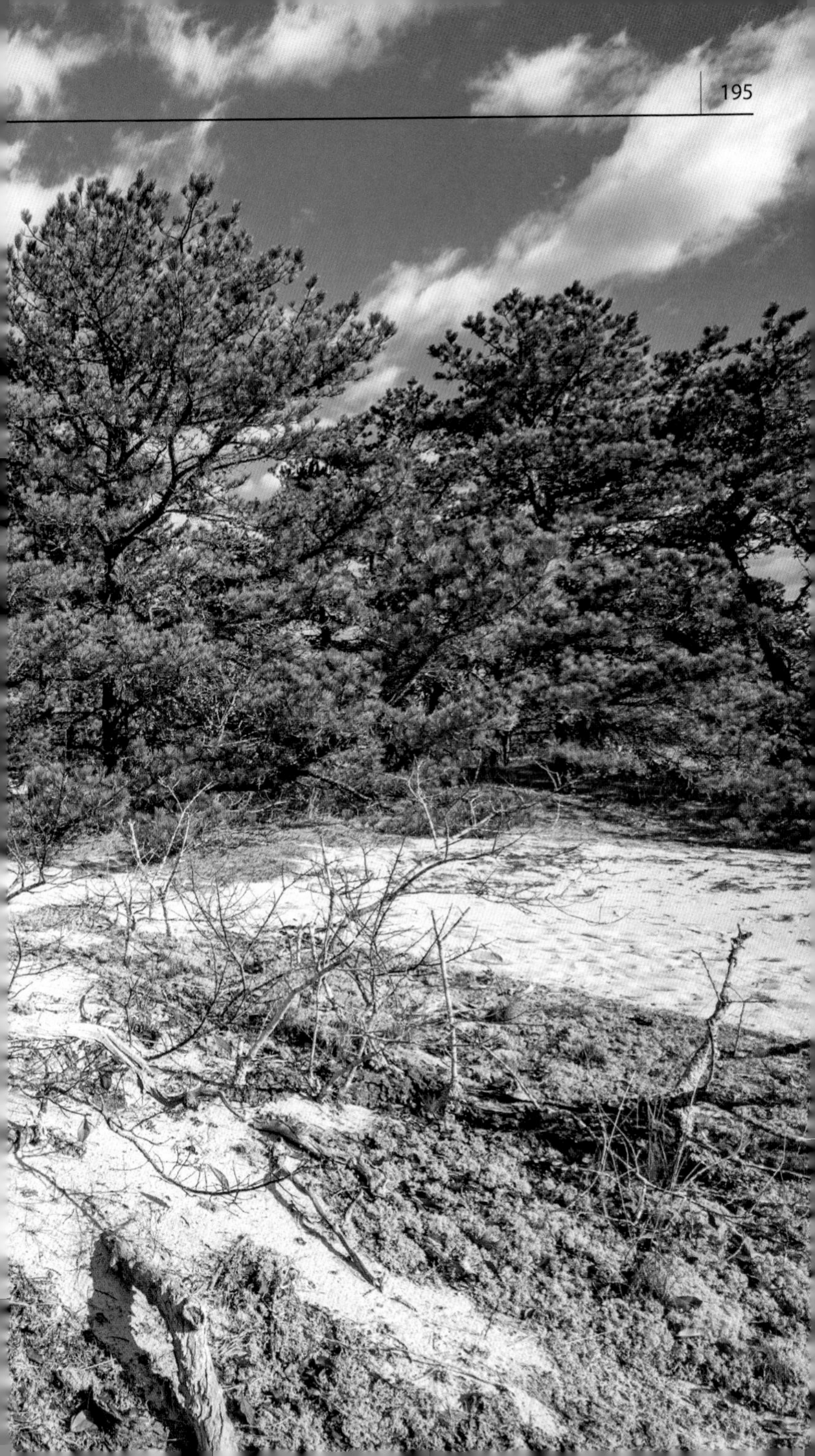

Salt Pond, Eastham, near the Visitor Center at Cape Cod National Seashore.

SALT MARSHES

High salt marsh meadows at Scorton Creek, East Sandwich, in June, when the marsh grasses are just beginning a season of lush growth.

Salt marshes are North America's most biologically productive ecosystem—only tropical rain forests and coastal mangrove ecosystems are their equals in productivity. Marshes provide significant nutrients to Outer Lands estuaries and are the coastal nursery ground for almost every important commercial and sport fish in nearshore waters. More than 170 fish species and 1,200 invertebrate species live in coastal New England waters at least part of the year, and most of those species use the salt marshes at some point in their life cycles.

Salt marshes are natural water treatment facilities, cleaning coastal waters through filtering by grasses and filter feeders that live in and around the marsh, as well as through the activities of the large detritivore community within marshes themselves. The natural salt marsh food chains have a large capacity to absorb and convert dissolved forms of organic matter into grass and animal biomass, cleaning coastal waters and adding vital nutrients to the Outer Lands ecosystem.

Marshes also act as sinks for the excess nitrogen that runs off our highly developed landscapes, and they absorb much of the excess carbon generated from the burning of fossil fuels. Salt marshes work as natural buffers and sponges in stormy conditions, protecting the coastline against storm surges and breaking the full force of storm waves and flooded rivers.

The productivity of salt marshes
Although few animals feed directly on marsh leaves, marsh grasses shed all their leaves during fall and winter, and this

major source of vegetable matter is broken down by bacteria and fungi into fine detritus that is carried out into the estuary by ebbing tides. Salt marsh grass provides a rich source of energy for tiny planktonic animals, filter feeders like clams, oysters, scallops, grass shrimp, amphipods, and other small animals. In turn, these small animals become food for larger predators and scavengers such as crabs, lobsters, fish, birds, and mammals.

Marshes around southern New England produce about 29 ounces per square yard of organic material each year, and that enormous productivity comes from three fundamental marsh components:

1. Mud algae, diatoms, and seaweeds at the marsh surface

2. Phytoplankton in the marsh water

3. Large salt marsh plants, especially grasses

Salt marshes are a unique form of grassland in that the entire above-ground annual growth dies back in winter, leaving only the underground rhizomes to renew the marsh in spring. Ninety percent of a marsh's annual productivity is realized at the end of the growing season in October, when the grass leaves begin to die off, decompose, and wash into the estuary through tidal flow over winter and early spring. In these tons of dead grass leaves per acre, much of the productivity drains from the marsh as an organic soup of plant detritus to be consumed by bacteria, fungi, and other tiny planktonic animals.

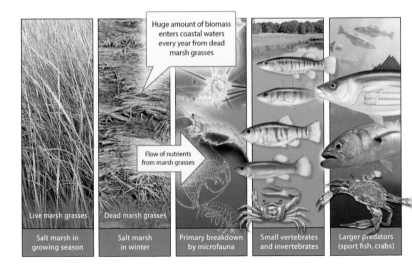

Huge amount of biomass enters coastal waters every year from dead marsh grasses

Flow of nutrients from marsh grasses

Live marsh grasses	Dead marsh grasses			
Salt marsh in growing season	Salt marsh in winter	Primary breakdown by microfauna	Small vertebrates and invertebrates	Larger predators (sport fish, crabs)

Comparative production rates

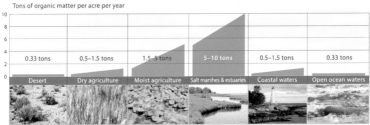

Tons of organic matter per acre per year

Desert	Dry agriculture	Moist agriculture	Salt marshes & estuaries	Coastal waters	Open ocean waters
0.33 tons	0.5–1.5 tons	1.5–5 tons	5–10 tons	0.5–1.5 tons	0.33 tons

After Teal and Teal, Life and Death of the Salt Marsh

Although these microscopic decomposer organisms are not easily visible to us, they are critical to the estuary food chain, because their activity makes the bounty of the salt marsh grasses available to the rest of the food chain.

Animals in the next link of the marsh food chain are detritivores such as fiddler crabs, salt marsh snails, grass shrimp, and marsh amphipods that consume the partially decomposed grass, turning it into animal biomass. Other partially decayed marsh detritus is swept out with each high tide into the estuary waters, where it is consumed by filter feeders such as mussels, barnacles, and clams.

Snails, clams, mussels, and other detritivores and filter feeders form the base of the food chain of predatory animals. Small marsh fish such as killifish and sticklebacks feed on detritivores like amphipods and grass shrimp, as do such crabs as the Green and Blue Crabs. The Diamondback Terrapin, crabs, and even such birds as the American Black Duck all feed on the abundance of salt marsh snails. Smaller marsh predators become food for larger animals such as Bluefish, Striped Bass, herons, and Ospreys. Although the productivity of the salt marsh may be hard to see directly, marshes are the largest contributors of biological wealth in our coastal waters and are the base of the food chain for most of the birds, fish, and other more visible wildlife along our shorelines.

Energy conversion
Salt marshes capture and convert about 6 percent of the sunlight that falls on them during the year. This figure may sound modest, but the salt marsh compares very well to other plant communities. A farm field of corn captures about 2 percent of the sun's energy; coral reefs capture about 3 percent. On average, an East Coast salt marsh creates 5–10 tons of organic matter per acre every year, with the warmer southeastern marshes at the higher end of the productivity scale. Most

Brian Gratwicke

Daggerblade Grass Shrimp
Palaemonetes pugio

heavily managed farm crops only produce about half that tonnage per acre per year.

Marsh ecosystems depend on tidal flow as well as rivers and
streams to deliver raw nutrients along with the mud and sediments washed from inland areas down to the coast. Because
marsh vegetation dies back each fall and winter, marshes
are quick to convert nutrients into organic matter available
to estuary and marine animals. In contrast to forests, where
much of the annual productivity is bound up in the wood and
roots of trees and other perennials for many years, marshes
release almost all of their annual production when the grasses
die back and are washed into the estuary to be broken down
and gradually converted into microscopic animal biomass at
the base of the food chain.

Regional salt marshes

Salt marshes on the shores of the Outer Lands are relatively
small because the coastline is so young relative to the far older
shores and huge salt marshes of the mid-Atlantic and southeastern coasts. New England has relatively few large areas of
tidal flats and coastal sediment beds that form the foundation
for salt marsh growth. Most current Outer Lands salt marshes
are less than 3,000 years old, and many are much younger.
The salt marshes that certainly existed on the shores of the
southern New England coastal plain up to about 25,000 years
ago were obliterated by the Laurentide Ice Sheet as it covered
the region. Although the ice sheet began to retreat 24,000
years ago, the slow recovery to temperate climactic conditions
and the rapidly rising sea level combined to limit the growth
of marshes in the region until relatively recently. Before that
time the ocean was rising at too fast a rate to permit the
long-term development of mature salt marshes. On the older
unglaciated coasts to the south the shoreline has had millions
of years to accumulate both the sandy barrier islands that
protect the shore from wave action and the rich, deep silt flats
that nurture the vast salt marshes of the southeastern Atlantic
coastline.

Most large salt marshes on the Atlantic Coast are associated
with major rivers or lie behind substantial barrier islands
formed at least in part from sediments borne to the coasts
by rivers. Because Cape Cod, Nantucket, Martha's Vineyard,
Block Island, and even New York's Long Island are too small
to provide the drainage for sizable rivers, most of the Outer
Lands salt marshes occur behind barrier islands, sandspits,
and very protected bays or along small, sluggish tidal rivers
and creeks. In sheltered waters such as those behind the
large Sandy Neck sandspit in Barnstable Harbor, silt from

coastal erosion can settle onto the bottom of tidal areas and eventually builds up to form a substantial platform for marsh grasses. As pioneer grasses like Saltwater Cordgrass (*Spartina alterniflora*) take root they trap yet more silt, building a higher platform of silt and peat from older grass roots, eventually forming high salt marsh dominated by Saltmeadow Cordgrass (*S. patens*), along with Spike Grass and Blackgrass.

Origins

Salt marshes typically form when seeds or rhizomes of Saltwater Cordgrass colonize shallow tidal flats. As the cordgrass shoots develop, they slow the movement of water, leading to sediment deposition and limiting erosion during storms and high tides. As the tidal flats grow into a more mature marsh platform, tidal creeks develop that drain the marsh during low tides. The movement of tidal water acts almost like a breathing mechanism for the marsh, bringing in fresh nutrients and sediment on the flood tide and draining away wastes and detritus on the ebb.

Marsh grasses propagate primarily by rhizomes—spreading underground stems that both expand the size and area of the original shoot of grass and help bind and stabilize the tidal mud beneath them. Rhizomes also allow marsh grasses to store energy underground for the next growing season.

As the grasses take root, the tangle of rhizomes and stems traps more sediments and sand, gradually building up a tough

The low marsh silt is held together by a complex of grass rhizomes, roots, Ribbed Mussels, and both macroalgae such as Sea Lettuce and microalgae on the mud surface.

Ribbed Mussels (*Geukensia demissa*) play an essential role in the low salt marsh, binding the marsh sediments to resist erosion from tides and storms and contributing nutrients to the *Spartina* through their feces.

platform of dense roots, covered by sticky surface mats of blue-green algae that are resistant to normal currents and tides. Once this platform reaches above the mean high tides, the marsh begins to stratify into a lower marsh area that floods twice a day with the tides and an upper marsh platform that floods only a few times a month during spring high tides.

Winter ice is a major limiting factor in northeastern salt marshes. Sharp, heavy plates of ice often cover the surface of the upper marsh and line the banks of marsh creeks in winter. These ice plates effectively shear off the stems of taller plants, so trees and bushes that might tolerate the salty water cannot gain a foothold in the upper marsh. The marsh grasses survive the ice shearing because their rhizomes and roots are safe under the mud and peat surface, ready to send up new green shoots in spring. Marsh environments experience a wide range of temperatures as well as rapid temperature changes, as the exposed marsh heats during low tides in warmer months and then cools rapidly when high tides flood the marsh.

Tidal movements, rainwater runoff, and the variable flow rate of rivers all affect the salinity of the water around marshes. Evaporation during low tides increases the salinity of shallow pools and open pannes in the salt marsh, sometimes to levels

High salt marsh (background) and a vibrant brackish marsh border (foreground) at Wellfleet Bay Wildlife Sanctuary.

In addition to edge and bramble species such as Fox Grape, Switch-grass, and goldenrods, there is a healthy line of Narrow-Leaved Cattails in the center foreground, a Groundsel Tree in bloom on the right, and an Eastern Black Oak on the far right, all common marsh edge species.

well above the salinity of ocean water.

Patterns and zonation

The structure of a salt marsh is a direct reflection of area tide levels. Its patterns of vertical zonation are so consistent that you can determine local tide levels simply by looking at the marsh plants, because each zone has characteristic vegetation. There are four major zones of the salt marsh:

Lower marsh

The area between the mean low water (MLW) line and the mean high water line (MHW). The lower marsh is dominated by a single grass species, Saltwater Cordgrass, which tolerates flooding at high tides but cannot thrive where its roots are permanently underwater.

Upper marsh

The marsh area above the MHW and below the mean spring high water line (MSHW), the highest of the monthly high tides. This area is dominated by Saltmeadow Cordgrass (Salt Hay), which can tolerate the twice-monthly flooding of spring high tides but otherwise must grow above the normal high tide level.

Narrow-Leaved Cattail
Typha angustifolia

The Scorton Creek salt marshes at Jones Lane in East Sandwich.

Salt pannes are low, muddy areas of the marsh with no grass cover. Pannes often flood during spring tides. The water they contain is often far saltier than seawater, owing to evaporation.

Salt pannes

Salt pannes are shallow, water-retaining, open areas within the marsh, often with a bare, muddy bottom and a sparse collection of plants. At low tide salt pannes may dry completely to a hard mud surface dotted with caked salt crystals. Older salt pannes sometimes develop a surface crust of algae and bacteria that will crack into plates during hot dry summer days. Though it isn't pretty, this living crust over the mud is an essential part of the biological productivity of the salt marsh. Mud Fiddler Crabs feed directly on the crust, and the tiny crabs often congregate around salt pannes.

Pannes are very high in salinity owing to the evaporation of brackish water. The thin vegetation of salt pannes is usually dominated by especially salt-tolerant plants, including Spike Grass, Glassworts, Sea Lavender, and sometimes a dwarfed, low-growing form of Saltwater Cordgrass.

Marsh borders

The higher ground surrounding the marsh, above the mean spring high water level, where flooding is rare and less salt-tolerant plants can survive. In most Outer Lands salt marshes you can spot the marsh border quickly by looking for Marsh Elder (High Tide Bush) and Groundsel Tree, two shrubs that normally line the salt marsh border.

The lower marsh

In southern New England the lower salt marsh vegetation is composed almost exclusively of Saltwater Cordgrass stands. Besides tolerating saltwater immersion of its roots and rhi-

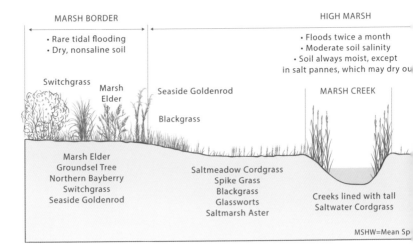

MARSH BORDER

HIGH MARSH

- Rare tidal flooding
- Dry, nonsaline soil

- Floods twice a month
- Moderate soil salinity
- Soil always moist, except in salt pannes, which may dry ou

Switchgrass

Marsh Elder

Seaside Goldenrod

MARSH CREEK

Blackgrass

Marsh Elder
Groundsel Tree
Northern Bayberry
Switchgrass
Seaside Goldenrod

Saltmeadow Cordgrass
Spike Grass
Blackgrass
Glassworts
Saltmarsh Aster

Creeks lined with tall
Saltwater Cordgrass

MSHW=Mean Sp

zomes, Saltwater Cordgrass has a range of other adaptations to living between the low and high tide lines. The waterlogged soils of salt marshes and tidal flats are poorly oxygenated, and all plant roots require oxygen to do the work of transporting nutrients throughout the plant. Saltwater Cordgrass has a special honeycomblike air circulation tissue called aerenchyma within its stems, rhizomes, and roots that allows the plant to grow on waterlogged soils and still oxygenate its roots. Its leaves also have special salt glands on their surface that excrete excess salt absorbed through its roots. If you look closely at a living cordgrass leaf, you'll see tiny white crystals of salt, excreted by the salt glands, all along the blade.

Although we normally look at the characteristic strengths of a dominant species like Saltwater Cordgrass in adapting to harsh conditions, it's always important to consider the competition a species faces from other plants. Saltwater Cordgrass is well adapted to life in marshes between the average low and high tide lines, but it is a poor competitor with grasses above the high tide line. The Saltmeadow Cordgrass (Salt Hay) that dominates the high marsh has tough, aggressive roots and rhizomes that prevent Saltwater Cordgrass from spreading into this zone. So in a sense Saltwater Cordgrass lives its life trapped between the low and high tide lines, well adapted to life there but unable to spread beyond that ecological niche.

A handful of other plant species can exist at the margins of the lower marsh. Sea Lavender will grow just below the high tide line, as will Glassworts. Spike Grass is the most salt tolerant of all the marsh grasses: it will grow in highly saline

Low salt marsh is defined by Saltwater Cordgrass (*Spartina alterniflora*), which grows especially tall where there is a big tidal range, as in Cape Cod Bay.

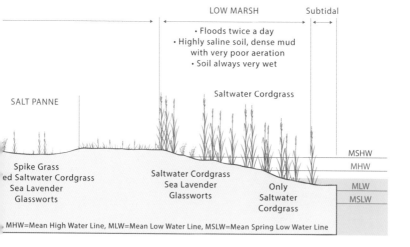

LOW MARSH — Subtidal

- Floods twice a day
- Highly saline soil, dense mud with very poor aeration
- Soil always very wet

Saltwater Cordgrass

SALT PANNE

Spike Grass
ed Saltwater Cordgrass
Sea Lavender
Glassworts

Saltwater Cordgrass
Sea Lavender
Glassworts

Only
Saltwater
Cordgrass

MSHW
MHW
MLW
MSLW

MHW=Mean High Water Line, MLW=Mean Low Water Line, MSLW=Mean Spring Low Water Line

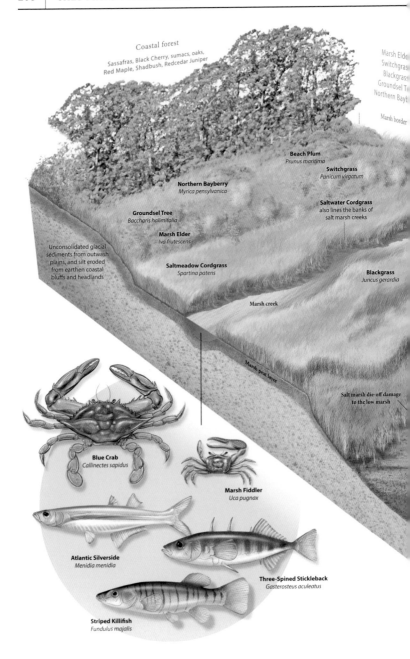

Coastal forest
Sassafras, Black Cherry, sumacs, oaks, Red Maple, Shadbush, Redcedar Juniper

Marsh Elde
Switchgras
Blackgrass
Groundsel Te
Northern Bayt

Marsh border

Beach Plum
Prunus maritima

Switchgrass
Panicum virgatum

Northern Bayberry
Myrica pensylvanica

Saltwater Cordgrass
also lines the banks of
salt marsh creeks

Groundsel Tree
Baccharis halimifolia

Marsh Elder
Iva frutescens

Unconsolidated glacial
sediments from outwash
plains, and silt eroded
from earthen coastal
bluffs and headlands

Saltmeadow Cordgrass
Spartina patens

Blackgrass
Juncus gerardia

Marsh creek

Marsh Peat layer

Salt marsh die-off damage
to the low marsh

Blue Crab
Callinectes sapidus

Marsh Fiddler
Uca pugnax

Atlantic Silverside
Menidia menidia

Three-Spined Stickleback
Gasterosteus aculeatus

Striped Killifish
Fundulus majalis

Habitat types, plant species, and animal species in the salt marsh are tightly correlated with tide heights. Just an inch or two of increased tide height can radically change the species mix in a marsh, one reason that sea level rise and climate change deeply worry biologists who study salt marshes and their inhabitants.

As sea levels rise we may lose many of our salt marshes—and the creatures that live in them or depend on them for food. We'll also lose the powerful benefits that marshes provide in protecting the coastlines during storms and in filtering pollutants that enter coastal waters from streams and rivers.

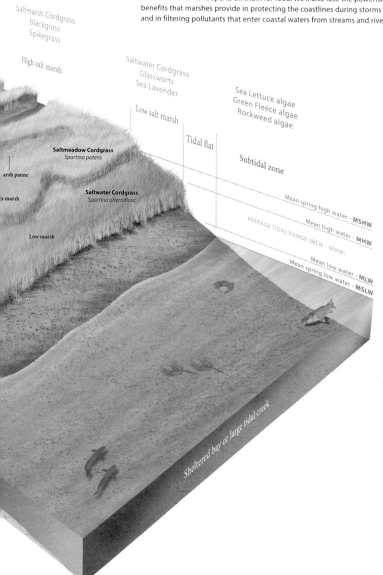

Saltmarsh Cordgrass
Blackgrass
Spikegrass

High salt marsh

Saltwater Cordgrass
Glassworts
Sea Lavender

Sea Lettuce algae
Green Fleece algae
Rockweed algae

Low salt marsh

Tidal flat

Saltmeadow Cordgrass
Spartina patens

Subtidal zone

arsh panne

n marsh

Saltwater Cordgrass
Spartina alterniflora

Low marsh

Mean spring high water - **MSHW**

AVERAGE TIDAL RANGE (MLW - MHW)   Mean high water - **MHW**

Mean low water - **MLW**
Mean spring low water - **MSLW**

Sheltered bay or large tidal creek

Sea Lavender (*Limonium carolinianum*) lends a touch of warm color to the greens and browns of the lower salt marsh. The flower stalks arise from a distinctive rosette of tough, leathery, dark green leaves.

Salt on a Saltwater Cordgrass (*Spartina alterniflora*) leaf.

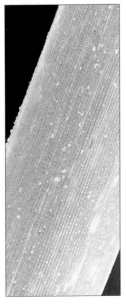

salt pannes within the high marsh but is sparse below the high tide line. A few large marine algae species are found in the lower salt marsh. Sea Lettuce is common in the lower marsh, as is the nonnative Green Fleece. Knotted Wrack and Rockweed will also grow in salt marshes if they can find stable attachment points. Often you'll see Rockweed and Knotted Wrack around drainage culverts and gates in the marsh or growing on large boulders on salt marsh creeks.

Under the lower marsh

The underlying silty soil in marshes is always soaked with water and is poor in oxygen. Anaerobic bacterial decomposition of organic matter in the soils leaves hydrogen sulfide as a by-product, giving marshes their characteristic rotten egg smell. The burrows of marsh animals such as fiddler crabs and marsh crabs aerate the marsh soils somewhat, but the anaerobic conditions prevent all but a few plants from taking root. The thick, sticky mat of blue-green algae, green algae, and bacteria on the surface of marsh mud is an important contributor to the overall biological productivity of the salt marsh. The algae mat also stabilizes the mud, prevents erosion, and traps new silt particles brought in with each new high tide, helping to build the marsh.

In the lower marsh, solid attachment areas are at a premium for such animals as Ribbed Mussels and barnacles, which attach to the bottom of grass stems and rhizomes of Saltwater Cordgrass. In this transition zone between the open water and creeks and the lower marsh, the byssal attachment threads and excretions of Ribbed Mussels help bind the loose silt sediments to the complex of mussel shells and Saltwater Cordgrass rhizomes, both protecting the lower marsh from washing away and slowly building the marsh by trapping sand and silt particles. The tight intertwining of Ribbed Mussels and Saltwater Cordgrass roots and rhizomes is mutually beneficial: the rhizomes give the mussels a firm anchor point in an environment that offers few places to attach, and the mussels increase soil nitrogen around the rhizomes, stimulating growth.

Much of the productivity of the lower salt marsh is created by blue-green and green algae, which attach to the stems of Saltwater Cordgrass and the mud surface in the lower marsh. If you look closely at the cordgrass stalks just above the water line or mud level, you will often see the tiny, coffee bean–like shells of salt marsh snails, which feed on the film of algae that grows on the base of cordgrass stalks. These air-breathing snails avoid immersion at high tide by climbing the grass stalks to stay above water level.

Invertebrates of the lower marsh and marsh creeks

In the intertidal zone of salt marshes you'll see many of the common species that also inhabit rocky shores, tidal flats, and beaches. The Eastern Mudsnail, Common Periwinkle, and Rough Periwinkle are all frequently found on mud flats and creek banks within the marsh. The snails eat algae and other organic material on the surface of the mudbanks and tidal flats.

On the creek mudbanks, usually near the top just under the upper layer of grasses, you may see groups of holes and small heaps of freshly dug mud. These are made by one of the two species of fiddler crabs that inhabit salt marshes. The constant digging of fiddler crabs is important for marsh grasses, because these burrows bring oxygenated water and nutrients into the silty soils. The most common fiddler crab in Outer Lands salt marshes is the Mud Fiddler Crab, but the Red-Jointed Fiddler Crab is also common. Both fiddler crabs feed on the rich layer of algae, bacteria, and plant detritus on the surface of marsh mud.

Glassworts (*Salicornia sp.*) often grow in the highly saline conditions of salt pannes.

Purple Marsh Crabs are about the size of a fiddler crab but are dark violet to black in color and have a more square-shaped body. These little crabs are less visible to the marsh visitor because they are active primarily at night. Although they will prey on fiddler crabs, Purple Marsh Crabs are primarily herbivores that feed on the stalks and leaves of Saltwater Cordgrass. In healthy lower marsh environments, the feeding of Purple Marsh Crabs has little effect on the abundance of Saltwater Cordgrass, but these crabs have been implicated in the die-off of lower marsh grasses in sections of many New England salt marshes. In marsh die-off syndrome, the crabs eat away large patches of Saltwater Cordgrass, leaving bare, muddy creek banks. Research by Brown University professor Mark Bertness and his graduate students has shown that marsh die-off tends to occur in areas where sport fishing has depleted the number of predatory fish and Blue Crabs that normally feed on Purple Marsh Crabs. Released from predation pressure, the enlarged crab population damages the salt marsh by eating far more grass.

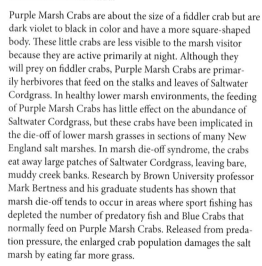

Within the marsh creeks the beautiful Blue Crab is the characteristic salt marsh crab. As the climate and water temperatures have warmed over the past few decades, Blue Crabs have been extending their range northward along the Atlantic Coast and have become more numerous in most areas. Blue Crabs are frequent targets of human crab fishers as well as the larger herons and so are wary of any movements around their area. If you are patient and stand still by a marsh creek

Mud Fiddler Crabs (*Uca pugnax*) are common in both the low and high salt marsh, where their burrows help aerate the dense, water-logged silt under the marsh surface.

SALTWATER CORDGRASS *Spartina alterniflora*

SALTWATER CORDGRASS *Spartina alterniflora*

SALTMEADOW CORDGRASS *Spartina patens*

SALTMEADOW CORDGRASS, cowlicks

High marsh ditch with both Saltwater Cordgrass (in ditch) and Saltmeadow Cordgrass (surrounding)

SALTWATER CORDGRASS AND RIBBED M

RSH ELDER *Iva frutescens*

RIBBED MUSSELS *Geukensia demissa*

KE GRASS *Distichlis spicata*

SWITCHGRASS *Panicum virgatum*

CKGRASS *Juncus gerardii*

BLACKGRASS, flower detail

GLASSWORTS *Salicornia sp.*

GROUNDSEL TREE *Baccharis halimifolia*

ERECT SEA BLIGHT *Suaeda linearis*

SEA LETTUCE *Ulva lactuca*

GREEN FLEECE *Codium fragile*

ROCKWEED *Fucus distichus*

LAVENDER *Limonium carolinianum*

SEA LAVENDER, basal rosette

SIDE GOLDENROD *Solidago sempervirens*

MARSH ORACH *Atriplex patula*

RENNIAL SALTMARSH ASTER *S. tenuifolium*

COMMON REED *Phragmites australis*

Salt marsh snails (*Melampus sp.*) on a Saltwater Cordgrass stem (*Spartina alterniflora*).

for a few minutes looking carefully under the surface, you will often be rewarded with spotting a Blue Crab or the smaller Green Crab, also common in salt marsh creeks.

Grass shrimp (also called prawns, *Palaemonetes sp.*) are also common in salt marsh creeks and play an important role in the salt marsh and estuary food chains. Grass shrimp are detritus feeders that break down larger bits of dead grass leaves. Bits of the reduced leaves that are not eaten by the shrimp become food for filter feeders like clams, mussels, and barnacles. The grass shrimp in turn are eaten by larger fish and birds that move into the marsh creeks at high tide.

Fish of the lower marsh and marsh creeks

Salt marshes are important nursery areas for many fish and are also rich in species that spend their lives in and around marshes. Small salt marsh fish eat algae, detritus from the breakdown of marsh grasses, amphipods, copepods, isopods, shrimp, marsh snails, and insects. Mummichogs and killifish are important predators on mosquito larvae, helping to limit populations of marsh mosquitos and other biting insects.

If you quickly approach a salt marsh creek in summer or fall, you'll see an explosion of tiny, panicked fish darting in every direction. The fish are small and fast-moving, and most are a nondescript brownish color, but if you wait until the fish settle down and look closely, you can identify a few species. The Common Mummichog is the most abundant fish in the lower marsh and marsh creeks. Mummichogs can grow as long as seven inches, but most individuals in marsh creeks and shallows are one to two inches long. Tiny flashes of silver are usually from the Atlantic Silverside, a fish that also ranges into the coastal and deeper waters of the Outer Lands. Striped Killifish and Sheepshead Minnows are other common residents of salt marsh creeks. Young Winter Flounder use the shelter of salt marsh creeks to grow before venturing into deeper waters, but the flounders are a favorite target for Ospreys, as are the schools of silvery Atlantic Menhaden and Blueback Herring that also enter tidal creeks in the marsh.

Larger species of coastal fish range into the marsh at high tide. Smaller Bluefish and Striped Bass enter for the rich pickings of small marsh fish and crabs in tidal creeks. These larger

Atlantic Silverside
Menidia menidia

predatory and sport fish species are important to salt marsh ecology because they limit the populations of plant-eating prey that might otherwise damage the marsh.

Diamondback Terrapin

Diamondback Terrapins are turtles native to brackish and salt marshes ranging from Cape Cod south to the Florida Keys and the Gulf Coast. Their common name derives from the diamond patterns of the shell carapace, the details of which are highly variable but are always in a bold geometric pattern. The name "terrapin" is derived from the Algonquian word "torope," which the Native Americans used to describe the Diamondback.

Terrapins are shy and are fast, strong swimmers with large webbed feet and powerful jaws for crushing their favored prey of small fish, clams, mussels, periwinkles, and mudsnails. To see them, approach salt marsh creeks slowly and scan the water for swimming turtles as well as the water's edge along the mudbanks for basking terrapins. In early summer, check roads near marshes for female turtles seeking out nesting sites. Terrapins can survive in the wild from 25 to 40 years, making them one of North America's longest-lived animals. Diamondbacks are also unusual in that they can survive in a variety of water salinities, from freshwater (<5 ppt salt) to ocean water (32 ppt), but they prefer the brackish water of salt marshes (15–25 ppt). Special lacrimal glands near their eyes allow terrapins to drink salt water and then excrete the salt as tears.

Diamondbacks mate in late spring and lay egg clutches in June and July. The females prefer to lay their eggs in sand-

A severe example of salt marsh die-off, at Stony Creek, Branford, Connecticut. Here almost all of the low salt marsh grass, Saltwater Cordgrass (*Spartina alterniflora*), has been chewed away by marsh crabs, leaving large brown banks of exposed marsh peat that erode easily in storms.

Diamondback Terrapin
Malaclemys terrapin

STRIPED KILLIFISH
Fundulus majalis

6–7 in.

THREE-SPINED STICKLEBACK
Gasterosteus aculeatus

2–4 in.

SHEEPSHEAD MINNOW
Cyprinodon variegatus

1.2–2.5 in.

COMMON MUMMICHOG
Fundulus heteroclitus

3–3.5 in.

**RED-JOINTED
FIDDLER CRAB**
Uca minax

Carapace
0.8 in. wide

**MUD FIDDLER
CRAB**
Uca pugnax

Carapace (shell)
0.7 in. wide

Carapace
1 in. wide

PURPLE MARSH CRAB
Sesarma reticulatum

BLUE CRAB
Callinectes sapidus

Carapace
7–8 in. wide

Color ranges from
bright green to gray-
green to brown

GREEN CRAB
Carcinus maenas

Carapace
3–3.5 in. wide

banks but will also dig nests under vegetation in the high marsh. Females often wander long distances from the marsh to find suitable nesting areas, making them vulnerable to cars and domestic animal predation. Young turtles emerge from eggs in August and September and are a favorite food of herons, Bluefish, and Striped Bass. Diamondbacks overwinter by hibernating in deep marsh creeks under mud bottoms or high marsh vegetation, but the Diamondback's winter biology is not well understood.

In the nineteenth and early twentieth centuries, the Diamondback Terrapin was nearly hunted to extinction for its meat, then used in a fashionable soup. As the popularity of turtle soup faded in the early twentieth century, southern New England populations of Diamondbacks recovered somewhat. Today the main threat to terrapins is habitat loss. Biologists estimate that almost 75 percent of terrapin marsh habitat has been eliminated since colonial times. Accidental death owing to human activity is another problem: terrapins are often caught and drowned in crab pots and nets or hit by boat propellers or cars. In Connecticut and New York the Diamondback Terrapin is not on either state's endangered or threatened lists, but terrapins are considered endangered in Rhode Island and threatened in Massachusetts.

Another common turtle that often enters salt marshes is the Common Snapping Turtle. Snapping Turtles are readily distinguished from terrapins by a larger head, lack of strong linear patterns on the shell, and generally darker overall color. Adult Snapping Turtles (up to 18 inches long) are also much larger than adult Diamondback Terrapins (usually five to

Diamondback Terrapin
Malaclemys terrapin

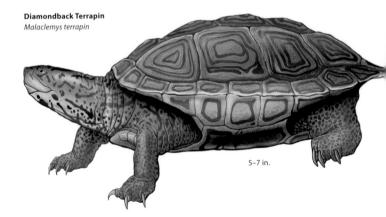

5–7 in.

seven inches long). Snapping Turtles are almost exclusively aquatic. They are more noticeable in late spring and early summer, when female turtles wander into the upper marsh or uplands near marshes to lay their eggs.

Birds of the lower marsh

The lower salt marsh offers a wide variety of animal and vegetal food sources to ducks, wading birds, and aerial divers like terns and Ospreys. Water is deeper in the low marsh, so long-legged wading birds such as the Great Blue Heron, Great Egret, and Snowy Egret are a common sight there in all seasons except winter—and there is enough food in salt marshes that some Great Blue Herons even hang around in winter. Wading shorebird species are other frequent marsh visitors, particularly during spring and fall migration. Greater and Lesser Yellowlegs and Glossy Ibis are very common in water less than a foot deep, and Willets both feed and nest in many marsh sites along the coastlines.

Short-Eared Owl
Asio flammeus

American Black Ducks and Gadwalls are year-round residents of Outer Lands salt marshes, and the habitat is an important breeding environment for these ducks, which have lost much of their former nesting habitat to coastal development. Other ducks use the marshes primarily in migration and in winter, when the marsh offers food as well as shelter from human activity and the weather. Given the relatively shallow water, the low marsh is used most intensively by dabbling duck species such as Mallards, Gadwalls, Green-Winged and Blue-Winged Teals, American Wigeons, and Wood Ducks. Deeper water near the marsh attracts the nonnative Mute Swan, as well as such diving ducks as the Hooded and Red-Breasted Mergansers, Greater and Lesser Scaups, Common Goldeneyes, and Ruddy Ducks. Double-Crested Cormorants also frequently dive for fish in or near salt marshes and can often be seen perched on docks, pilings, and other nearby structures.

Wood Duck
Aix sponsa

Coastal diving birds like terns depend on the health of our salt marshes. Terns feed on small fish such as Common Mummichogs, Atlantic Silversides, Sand Lance, and Sheepshead Minnows, which they take in shallow dives at the water surface. All of these small fish either live entirely in or around salt marshes or depend on the protection of marshes early in their lives. Common Terns nest primarily in colonies on offshore islands like Monomoy Island south of Chatham Island and feed all along the coast and in the central parts of Nantucket Sound and Cape Cod Bay, but their prey fish are all nurtured in salt marshes. The endangered Least Tern takes small fish directly from salt marsh creeks and from shallow coastal waters near marshes. Black Skimmers (more frequent

Green-Winged Teal
Anas crecca

Common Snapping Turtle
Chelydra serpentina

on Long Island) also hunt more open waters near marshes for the same small fish species taken by terns.

Along creek banks in the lower marsh look for the secretive Clapper Rail in early morning or evening twilight hours. Scanning the same areas of marsh creeks will often turn up a Black-Crowned Night-Heron or the less common but similar Yellow-Crowned Night-Heron. Yellow-Crowned Night-Herons are uncommon on Cape Cod, Nantucket, and Martha's Vineyard but are frequently seen in coastal Rhode Island and eastern Long Island. All three birds hunt for fiddler crabs, snails, and other small marsh invertebrates. The night-herons will also take larger crabs and fish from the marsh creeks.

The upper marsh

The upper marsh, or marsh platform, is the area of salt marshes above the typical high tide line (MHW) but below the level of the high water in monthly spring tides (MSHW) (see illustration, p. 227). Although the upper marsh is above the average high tide, it is flooded twice monthly by spring high tides, and thus the plants in the upper marsh must also be able to tolerate regular immersions in salt water. As with the lower salt marsh, the upper marsh owes its existence to a single dominant grass species, in this case Saltmeadow Cordgrass. Saltmeadow Cordgrass (Salt Hay) is the low-growing grass that forms the open meadowlike expanses most people think of when referring to salt marshes. It is brilliantly green

American Black Duck
Anas rubripes

from late spring to early fall and has a peculiar growth habit of not usually growing fully upright. The grass stalks tend to lean over against their neighbors, giving Saltmeadow Cordgrass meadows their typical cowlicked appearance.

In northeastern salt marshes Saltmeadow Cordgrass is joined by scatterings of Spike Grass. Spike Grass is not competitive enough with Saltmeadow Cordgrass to dominate most marsh areas, but in areas of the upper marsh where the soil has been disturbed or is particularly salty, Spike Grass may occur in pure stands. Spike Grass is the most salt tolerant of all high marsh grasses and will also grow in salt pannes in the upper marsh, where because of evaporation the underlying mud is often much saltier than pure seawater. In typical mixed marshes Spike Grass is not easy to spot among the more common Saltmeadow Cordgrass, but from September on, the distinctive white flowers of Spike Grass are visible, making the overall distribution of Spike Grass in the marsh more obvious.

A classic salt meadow composed of Saltmeadow Cordgrass (*Spartina patens*).

In raised areas of the salt marsh or along the upland rim of the marsh, Blackgrass (Black Rush) mixes in with Saltmeadow Cordgrass and Spike Grass. Areas of the marsh dominated by Blackgrass have a different visual texture and color, because Blackgrass leaves are spiky and erect as well as a darker shade of green in summer. In the spring Blackgrass is the first marsh grass to turn a brilliant spring green, and in the fall Blackgrass leaves turn the very dark brown or black color that gives the rush its name.

ack-Crowned
ight-Heron in flight

Yellow-Crowned
Night-Heron in flight

**Yellow-Crowned
Night-Heron**
Nyctanassa violacea

**Black-Crowned
Night-Heron**
Nycticorax nycticorax

GREAT BLUE HERON *Ardea herodias*

GREAT EGRET *Ardea albus*

SNOWY EGRET *Egretta thula*

WILLET *Tringa semipalmata*

LESSER YELLOWLEGS *Tringa flavipes*

Glenn Young

GREATER YELLOWLEGS *Tringa melanoleuca*

SSY IBIS *Plegadis falcinellus*

MALLARD *Anas platyrhynchos*

Byland

E-WINGED TEAL *Anas discors*

Steve Oehlenschlager

GADWALL *Anas strepera*

E SWAN *Cygnus olor*

pstclair

CLAPPER RAIL *Rallus longirostris*

Common Salt Marsh Border Shrubs

Marsh Elder
Iva frutescens

Groundsel Tree
Baccharis halimifolia

Northern Bayberry
Myrica pensylvanica

Beach Plum
Prunus maritima

Black Cherry
Prunus serotina

In the less salty conditions of the upper marsh more plants are able to tolerate the occasional baths of salt water. Two herbaceous plants are very common in the upper marsh: Glassworts and Sea Lavender. Glassworts have a distinctive twig and leaf structure adapted to conserve water and resist salt spray. The leaves are much reduced and hug the fleshy stems, but Glassworts are true flowering plants. Unless you are a botanist the three Glasswort species are difficult to distinguish, and botanists themselves do not always agree on which is which, so we'll just call them Glassworts. Luckily, Glassworts are so distinctive that once you know what they look like you can easily spot them in the marsh, particularly in the fall, when Glasswort stems turn a brilliant red against the green of the marsh grasses around them. When in bloom in the summer and early fall, Sea Lavender adds a beautiful violet haze to the waterside edges of the upper marsh. Sea Lavender is particularly tolerant of salt water on its roots and will even grow slightly below the high tide line.

Scattered through the upper marsh grasses you'll see a number of other flowering plants. In the late summer and fall, the golden flowers of Seaside Goldenrod are unmistakable, and even before they flower, the fleshy, tough leaves of this perennial stand out in the upper sections of salt marshes. The much less conspicuous Common Orach and Perennial Saltmarsh Aster are also common along the upland rim of the high marsh.

Invertebrates of the upper salt marsh

As in the lower marsh, the most visible invertebrates of the upper marsh are the small Purple Marsh Crabs, Mud Fiddler Crabs, and Red-Jointed Fiddler Crabs that scurry into holes or under grasses before you as you walk through the marsh. However, by sheer number the dominant invertebrate of the upper marsh is the salt marsh snail (*Melampus sp.*), sometimes called the coffee bean snail because of its size and glossy brown color. These tiny snails are the most abundant invertebrate in the high marsh, occurring in densities of hundreds per square yard in healthy marshes. Salt marsh snails start their lives as aquatic larvae in nearby estuary waters, but as adults the snails are air-breathing and avoid immersion in water by climbing grass stems during high tides. Salt marsh snails are an important link in the salt marsh food chain. They feed on algae and grass debris on the surface of the marsh and in turn are eaten by American Black Ducks, Diamondback Terrapins, and other larger marsh animals.

Inspection of the grass and underlying mud surface will show other upper marsh invertebrates, although not in the same

e critical role of tides in salt marsh formation

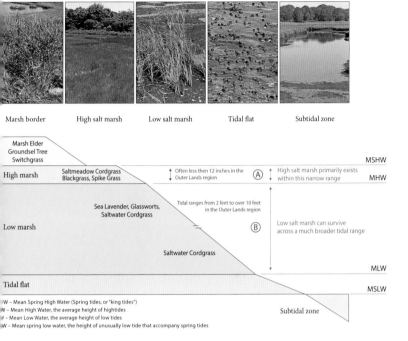

| Marsh border | High salt marsh | Low salt marsh | Tidal flat | Subtidal zone |

| Marsh Elder Groundsel Tree Switchgrass | | | | | MSHW |
| High marsh | Saltmeadow Cordgrass Blackgrass, Spike Grass | ↑ Often less then 12 inches in the ↓ Outer Lands region | (A) ↑ High salt marsh primarily exists within this narrow range ↓ | MHW |

Sea Lavender, Glassworts, Saltwater Cordgrass

Tidal ranges from 2 feet to over 10 feet in the Outer Lands region

(B)   Low salt marsh can survive across a much broader tidal range

Low marsh

Saltwater Cordgrass

MLW

Tidal flat

MSLW

W – Mean Spring High Water (Spring tides, or "king tides")
W – Mean High Water, the average height of hightides
– Mean Low Water, the average height of low tides
W – Mean spring low water, the height of unusually low tide that accompany spring tides

Subtidal zone

numbers as fiddler crabs and salt marsh snails. Salt marsh isopods and amphipods are small, pillbug-like crustaceans that feed on decomposing salt marsh grasses and algae from the surface of the marsh soils. Wolf spiders are a common upper marsh predator of small insects. You won't have to look for one unfortunately common insect—if they are present in the marsh, the biting greenhead flies will find you.

The various species of salt marsh mosquitos are other pests you'll come across in the summer and fall months. It's always wise to bring a DEET-based repellent along for any trip into a salt marsh in summer or early fall, both to avoid the annoyance of insect bites and to prevent more serious problems that can come from insect- or tick-borne diseases. Marsh mosquitos can carry West Nile virus, although the chances of catching the disease from the average mosquito bite is extremely low. A much more serious problem is the presence of ticks. Even if it's 90 degrees Fahrenheit on a sunny summer day, never "bushwhack" through marsh vegetation with bare legs, even if you are wearing insect repellent. If you plan on straying off established paths, always wear long pants sprayed with

High salt marshes are in critical danger from today's rapid sea level rise because high marsh exists in the very narrow range between the average height of high tides and the average height of twice-monthly spring tides. The most important high marsh grass, Saltmeadow Cordgrass (*Spartina patens*), cannot tolerate more than the twice-monthly soaking of its roots in salt water. Even a few millimeters of sea level rise can begin to damage and eventually kill high salt marsh meadows.

Osprey
Pandion haliaetus

The return of the Osprey

These days the most emblematic bird of the upper salt marsh is the Osprey, which both nests and hunts over salt marshes throughout the Outer Lands. The return of the Osprey as a common coastal bird is a wonderful environmental success story.

After World War II the widespread use of the insecticide DDT devastated American populations of Ospreys, because DDT and its organochlorine breakdown products readily entered the coastal food chain and became concentrated in top-level predators such as the Osprey. The DDT-based chemicals made the eggshells of birds of prey like the Osprey too thin to hatch successfully, and in the postwar decades Osprey populations plunged in the northeastern United States.

After the use of DDT was banned in the United States in 1972, New England's Osprey population began a long, slow recovery. Today the Osprey is once again one of our most visible coastal hawks, and Osprey nests are a common sight above salt marshes.

Lone Star Tick
Amblyomma americanum

Black-Legged Tick
(often called a deer tick)
Ixodes scapularis

Carrier of Lyme disease

American Dog Tick
Dermacentor variabilis

greenhead fly
Tabanus sp.

Photos: Melinda Fawver, Sarah2, photobee, Elliotte Rusty Harold, allocricetulus.

a DEET-based repellent. There are just too many shrubs and long grass stems in the marsh to risk wearing shorts, and long pants will also protect you from other common problems like Poison Ivy in marsh border areas. Both the Black-Legged Tick and the American Dog Tick are common in salt marshes. The Black-Legged Tick is the vector for the Lyme disease spiro-chete bacteria *Borrelia burgdorferi*. American Dog Ticks can carry diseases like Rocky Mountain spotted fever, but luckily that disease is rare in the Northeast.

Birds of the upper marsh

In spring and fall migration and over the winter, Northern Harriers (formerly called Marsh Hawks) and Short-Eared Owls hunt over the Outer Lands salt marshes. Sadly, in recent decades the Short-Eared Owl population has been sharply reduced owing to the loss of both freshwater and saltwater marsh habitats, and today the sight of the Short-Eared Owl's low, tilting flight over the marsh is an unusual moment to treasure. Northern Harriers tell a happier story. Once devastated by the same DDT eggshell problems as the Osprey, harriers are now a common sight all along the coast in every season except the height of summer. The nearly ubiquitous Red-Tailed Hawk commonly hunts Meadow Voles in the salt marsh, and you'll often see American Kestrels perched on dead snags with a good view of the marsh or hovering over the marsh hunting for their prey of larger insects and small mammals.

The larger heron species are usually the most visible birds within the upper marsh, hunting along natural tidal creeks and the straight, artificial mosquito ditches for crabs and small fish. The Great Blue Heron, Great Egret, and Snowy Egret are the most often spotted, but the smaller Green Heron is also common, if less visible owing to its size and more secretive habits. The even shyer Clapper Rail both feeds and breeds in the upper marsh but is rarely seen because of its retiring nature and activity at dawn and dusk. Black-Crowned Night-Herons prowl marshes at twilight but also commonly fly and feed during daylight hours. Willets often nest in or near the upper marsh, and if you are near a nest, the Willet pair will circle you, calling loudly and displaying their bright white wing stripes as they fly or land nearby. If this happens, please retreat from the area, because a close approach to the nest stresses both the adults and eggs or nestlings.

Our common gull species also frequent salt marshes in search of food. Besides the birds themselves, you will often see the footprints of Herring and Ring-Billed Gulls in salt pannes and creek banks of salt marshes, where the gulls particularly relish

Steve Byland

eating fiddler crabs and the larger crab species.

Small songbirds are a major part of the bird life of the upper marsh. The Seaside Sparrow and the Saltmarsh Sparrow are salt marsh specialists and are rarely seen in other environments. These marsh sparrow species cling precariously to the tops of marsh grasses while singing their territorial songs but can also behave almost more like mice than birds, running head-down and low through grasses to avoid detection. Marsh Wrens are the third of the classic salt marsh songbird species around the Outer Lands and are common and noisy residents of both saltwater and freshwater marshes.

Northern Harriers (*Circus cyaneus,* also called Marsh Hawks) are a common sight over northeastern salt marshes. The distinctive owl-like round face of the harrier is an adaptation to hunting by ear. Harriers sweep over the marsh barely above the tops of the grasses, hoping to surprise a Meadow Vole feeding out in the open.

Mammals of the upper marsh

The upper platform of salt marshes is a dry enough habitat to attract mammals, at least as a food resource, and a few smaller mammals live in the marsh itself. Although you won't often see them, Meadow Voles are common in the upper marsh, and it is these voles that attract Northern Harriers, Red-Tailed Hawks, and other predatory birds to the upper salt marsh. Muskrats are common in brackish salt marshes that transition to freshwater wetlands. Raccoons are clever and versatile omnivores that frequently enter the upper salt marsh in search of mussels, crabs, and other small animals. Raccoons are chiefly nocturnal, so you won't often see them, but you can find their tracks in salt pannes or creek banks in the marsh.

Meadow Vole
Microtus pennsylvanicus

Although the Eastern Cottontail rarely grazes in the upper marsh itself, rabbits are very com-

A small corner of the vast salt marshes behind Sandy Neck, Barnstable.

Willets (*Tringa semipalmata*) commonly nest within or near salt marshes. If a pair of Willets like this one circles over you, calling noisily, try to back off and choose another path if you can. The birds are protecting a nearby nest.

mon in coastal parklands with grassy areas next to salt marshes.

The upper borders of the salt marsh

Four dominant border plants allow you to see the marsh border above the maximum high tide line. Marsh Elder is the classic high tide bush and the most visible marker for the salt marsh border. Once you can identify Marsh Elder, you will be able to read the marsh quickly, because Marsh Elder typically grows in a narrow band right up against the mean spring high water line, where it must tolerate some salt water in storms. But Marsh Elder is not competitive enough with more terrestrial shrubs and grasses to spread far beyond the MSHW line. Switchgrass joins Marsh Elder along the marsh border but also spreads into more upland areas along the marsh border, and Switchgrass is also common in other coastal environments like upper beaches. Groundsel Tree is another shrub that grows near the MSHW line, but it is usually found on ground a bit higher and farther back from the marsh edge. The fourth highly visible marsh edge plant is Seaside Goldenrod, which is sparse within the high marsh but becomes very common in marsh border areas.

The invasion of *Phragmites*

Along virtually all New England and Long Island salt marsh upper borders there are stands of the invasive Eurasian subspecies of the Common Reed (*Phragmites australis australis*), often simply called *Phragmites*. *Phragmites* is particularly well adapted to disturbed ground that is a little saltier than average, although this reed species is so adaptable that it can

SWAMP SPARROW
Melospiza georgiana

SONG SPARROW
Melospiza melodia

SEASIDE SPARROW
Ammodramus maritimus

MARSH WREN
Cistothorus palustris

SALTMARSH SPARROW
Ammodramus caudacutus

NELSON'S SPARROW
Ammodramus nelsoni

The Saltmarsh Sparrow is one of North America's most endangered species, owing to the loss of high marsh areas through sea level rise and habitat destruction.

The Green Heron (*Butorides virescens***)** is the smallest of the herons in Outer Lands marshes. This heron can be either stealthy and inconspicuous or remarkably tame and approachable. Either way, the Green Heron is a versatile predator of large insects, small fish, frogs, and worms and is common in both freshwater and saltwater wetlands.

live in just about any kind of wetland area except true salt marshes, where it inhabits the marsh border.

Biologists differ on the ecological value of *Phragmites* as a food source. For decades the environmentalist's view of *Phragmites* was that it was a cancer on the landscape, driving out native plants while supplying little nutritive value to wetland ecosystems. More recent research on *Phragmites*' impact has provided a more balanced view, showing that *Phragmites* does contribute useful primary productivity and biomass to coastal and wetland ecosystems, albeit not to the same degree that the displaced native plants like the Narrow-Leaved Cattail formerly provided.

Salt marsh conservation

Thanks largely to progress through the modern environmental movement, there is now a broader societal understanding of the ecological importance of wetlands and the economic and practical value of natural coastal habitats. However, there isn't a square inch of Outer Lands shoreline that hasn't been heavily influenced or thoroughly modified by human activity, and over the past 300 years, more than half of the region's salt marsh habitat has been lost to filling and development. The story in our region is not unique—there were approximately 220 million acres of salt marsh in North America in precontact times, and today only about 104 million acres of marshes remain intact.

In the past 40 years the importance of protecting both inland and coastal wetlands has driven both state and federal legal protections for wetland areas, but coastal salt marshes continue to face a host of threats beyond potential burial under new shoreline construction projects. Owing to both long-term climate change and the more recent acceleration of global warming and sea level rise, salt marshes in southern New England and on Long Island are now in a dangerous period where the accelerating rise in sea level may be too fast for our existing marshes to adapt.

In many locations along the New England and Long Island coast you can see the remains of former salt marshes along what are now becoming beaches, rocky shores, or tidal flats. Large, dark chunks of salt marsh peat, the remains of former marshes, lie surrounded by new sand and mud; the peat gradually washes away in storms, and all traces of the former marsh vanish.

In the Outer Lands region the sea level has risen about 18 inches over the past 300 years. The recent hurricanes Irene and Sandy tore away large chunks of salt marsh in Long Island's Peconic Bay, in the Sandy Neck marshes of Barnstable Harbor, and in the Eastham and Wellfleet salt marshes. Most smaller salt marshes are situated in more protected bays and inlets and suffered less damage in recent storms, but all of these salt marshes face the same ultimate challenge—the

Eastern Cottontail
Sylvilagus floridanus

A shallow salt marsh panne with stunted Spike Grass (*Distichlis spicata*), Glassworts (*Salicornia sp.*), and Saltwater Cordgrass (*Spartina alterniflora*). The highly saline water in pannes, plus the harsh temperature regime of hot, dry low tides followed by cold, wet high tides, stunts the few plants that can survive in salt pannes.

marshes are ringed by roads and houses, and as the sea rises, there is no room for the marshes to retreat from the higher water.

Change comes slowly in salt marshes. The mosquito control ditches that were cut into most of the region's larger salt marshes in the mid-1930s remain open after 80 years. The ditching accomplished little to control mosquitos, but the ditches exist to this day as reminders of how even such a productive ecosystem as the salt marsh may not be able to repair itself and adapt within human time scales.

Eastern Redcedar Junipers (*Juniperus virginiana*) are one of the most common trees along the borders of salt marshes and at the seaward edges of maritime forests.

Salt marshes are critical habitats for many of the Cape's most visible waterbirds. A Great Eget (*Ardea alba*) lifts off from a marsh creek in Eastham.

CK CHERRY *Prunus serotina*

COMMON JUNIPER *Juniperus communis*

IGED SUMAC *Rhus copallina*

STAGHORN SUMAC *Rhus typhina*

RTHERN BAYBERRY *Myrica pensylvanica*

BEACH PLUM *Prunus maritima*

A lone dwarfed Pitch Pine stands within a heath community in the Marconi Station area of Cape Cod National Seashore in Wellfleet. The hillside beyond the pine is classic coastal heath.

Heaths, Pine Barrens, and Grasslands

*Yellow Lance-Leafed Coreopsis (*Coreopsis lanceolata*) and purple Cow Vetch (*Vicia cracca*) in the grasslands around Highland Light, North Truro, part of Cape Cod National Seashore.*

The coastal heaths, pine barrens, and grasslands may be the first images that spring to mind when people envision Cape Cod and the islands, and in particular Cape Cod National Seashore, where mile after mile of low, rolling hills is carpeted in heath vegetation, dotted with scattered Pitch Pines dwarfed and twisted by sea winds, and breakers roar on the distant beaches 100 feet below the cliff rim of the Outer Cape. It's a gorgeous scene on a sunny summer day, but like many Outer Lands landscapes, this wonderfully wild and beautiful view has been created by a complex past dominated by human activity.

In the mid-1800s, after two centuries of clear-cutting, poor agricultural practices, and grazing by free-roaming pigs, sheep, and cattle, the high outwash plains of the Outer Cape between Eastham and Truro were barren ruins. Many Mid-Cape towns were in scarcely better condition. Most farms were abandoned or barely hanging on, the forests had long since vanished, and the people of Cape Cod had turned to fishing, whaling, and crewing sailing ships to earn their living. Even as late as the mid-1920s Henry Beston described his walk between what is now Coast Guard Beach in Eastham and the Salt Pond area around the National Seashore Visitor Center as "desolate and half desert," "a belt of wild, rolling, and treeless sand moorland," covered mostly by carpets of Beach Heather—also called Poverty Grass for its ability to survive on poor soils. While Beston also called the plains area "extraordinary beautiful," he would scarcely recognize the same neighborhood today, with its thick forest mix of Pitch Pines and scattered oaks.

Succession in wild communities

Grasslands, heaths (sometimes called moors), and some pine barrens are what biologists call early succession environments, meaning that they are wild habitats in transition to another type of habitat. Succession is a familiar concept to most of us because we see it happening all the time. In most of southern New England a bare patch of ground left to grow wild would typically succeed over time from grasses and herbs to grasses, herbs, and shrubs, then to shrubs and small trees, and finally, given decades of growth, to mature hardwood forest. Depending on such environmental factors as rainfall, soil moisture and minerals, average temperatures, and exposure to coastal salt spray, different mixtures of plant and animal communities will develop in succession. In most of southeastern coastal New England the culmination of natural succession is a climax community of mixed pine and hardwood forest, which once developed will persist relatively unchanged until fire or other disaster damages the trees.

Heath grades into sparse pine barrens forest in the Marconi Station area of Cape Cod National Seashore. Given a few more decades the Pitch Pines will probably flood the area and it will become true pine barrens forest. The lighter green ground cover is Bearberry (*Arctostaphylos uva-ursi*), interspersed with darker green mounds of Broom Crowberry (*Corema conradii*). A small Bear Oak (center left) and a Northern Bayberry bush (far right) complete the cast of standard heath characters.

Before European settlers and farmers clear-cut the land and developed agriculture, the Outer Cape was mostly a mix of mature climax forests of Pitch Pines and oaks of various species. In smaller, very sandy coastal areas with naturally poorer and drier soils, exposed to salt spray, the climax community was a particular mixture of ground covers, shrubs, and Pitch Pines called a pine barrens community. Grasslands and heaths also existed in the Outer Lands before European settlement, but those open environments were unusual and often

temporary, caused by natural wildfires or fires deliberately set by Native Americans to clear land or by blow-down damage from hurricanes and other severe storms. In some dry, sandy coastal areas exposed to heavy salt spray, heath was the climax community, but in precontact times those climax heath communities were probably small and confined to ocean-facing coasts in eastern Long Island, Block Island, Martha's Vineyard, Nantucket, and the Outer Cape highlands.

Today's wild communities in the Outer Lands are vastly different from those that existed before the Pilgrims landed in 1620, mostly because 400 years of human activity have affected virtually every square foot of the region. In the Outer Cape, Nan-tucket, southern Martha's Vineyard, and in many places on the ocean

Yellow-Rumped Warblers (*Setophaga coronata*) are one of the few bird species adapted to feed on the waxy berries of coastal heath and dune plants like the Northern Bayberry (*Myrica pensylvanica*).

The heaths of Marconi Station in Wellfleet, part of Cape Cod National Seashore.

Large stretches of coastal heath and grasslands are rare in the Outer Lands, but smaller areas of both habitats are fortunately more common. At the Wellfleet Bay Wildlife Sanctuary there are small heath and grassland areas in the uplands overlooking the salt marshes. The gorgeous Wing Island Trail at the Cape Cod Museum of Natural History in Brewster also features sections of heath and grassland.

coasts of Long Island heaths, grasslands, and pine barrens are now more common, and the Outer Lands in total are more ecologically diverse than they were in 1620. But in most cases those diverse coastal habitats are early succession communities, gradually reverting back to climax communities of coastal forest. This poses a complex dilemma for today's managers of wild lands: Should attempts be made to maintain the current diversity of grasslands, heaths, and pine barrens, or should nature be permitted to revert in most areas to coastal forest? The park managers of Cape Cod Natural Seashore have maintained a complex balance of approaches, allowing forest to redevelop in some areas but actively managing other areas to maintain grasslands and heaths.

Heaths

Heaths are a community of low-growing ground cover plants and shrubs that typically form on sandy, acidic, fast-draining soils that are poor in nutrients. The largest heaths in the Outer Lands occur on the tops of the ocean-facing cliffs of Cape Cod, where constant salt spray from the ocean tends to dwarf or kill typical forest trees. Heaths dominate the cliff areas starting in the Easton area, extending north to Marconi Beach and Marconi Station areas, and then farther north to the Highland Light area in Truro. In the Outer Lands heaths there are typically a scattering of small, mostly shrublike Pitch Pines and Bear Oaks, often dwarfed by the dry conditions and salt spray. The dominant plants in heaths are low ground covers including Bearberry, Broom Crowberry, Beach Heather (Poverty Grass), Sweet Fern (not a true fern), Wintergreen

(Teaberry), Lowbush Blueberry, Dusty Miller, and Seaside Goldenrod. In early summer you may see bright sprays of Yellow Wild Indigo. Nonflowering plants are common. Gray Reindeer Lichen forms rounded gray mounds mixed into the Bearberry and Broom Crowberry, and the branches of shrubs and small trees are often covered with Smooth Beard and Bushy Beard Lichens (Old Man's Beard).

Small Pitch Pines occur in isolation or in small groups, along with such shrubs or shrublike coastal trees as Bear Oak, Northern Bayberry, Highbush Blueberry, Beach Plum, Black Cherry, and scattered dwarfed specimens of White, Black, and Scarlet Oaks. Dwarf Chinkapin Oak also mixes into heaths, but not in large numbers.

The best place on Cape Cod for an introduction to heaths is the Marconi Station area of Cape Cod National Seashore in Wellfleet. The road to the remains of Marconi's now-vanished radio broadcasting station is lined with beautiful heath as well as heath grading into sparse pine barrens. Several unmarked but public roads and hiking trails allow you to pull over and walk into the heaths and pine barrens for a closer look. You may see open areas where the park managers have burned

Classic coastal heath along the Bearberry Hill Trail, above the eastern end of North Pamet Road in Truro. Here Bearberry, Broom Crowberry, and Atlantic Beachgrass ground covers mix with low Northern Bayberry, Bear Oak, and Highbush Blueberry bushes. A single dwarfed Pitch Pine (top left) shows rusty red damage from salt spray, even though the beach is 50 feet below the cliff edge behind.

Northern Saw-Whet Owl
Aegolius acadicus

Saw-Whet Owls are one of North America's tiniest owl species, and in fall and winter they frequently nest and feed in dense stands of coastal conifers, including pine barrens. Roosting Saw-Whets can be remarkably tame, and birders who locate them are sometimes tempted to move in close for photos. Wild birds in winter are always on the knife-edge of food availability versus their own calorie consumption, and it is very unhelpful to cause a resting bird to flush and waste precious energy. Appreciate your good luck in finding the bird, but keep your distance.

over and cut out the trees in an effort to keep the heath from reverting entirely to Pitch Pine forest. The high platform at Marconi Station affords a sweeping view of the heath and pine barrens–covered plains of the Outer Cape.

Animals of the heaths

A number of insect species are heath specialists, including the unusual moth Gerhard's Underwing. In spring and summer large numbers of Organ Pipe Mud Dauber wasps course low over the health vegetation, hunting spiders. Bright green tiger beetles are an unusual but interesting find. Exposed patches of sandy soil are usually dotted with anthills, and those millions of ants attract antlions. Study the sand for the reverse of an anthill: a steep-sided cone pit in the soil is a trap for unlucky ants that blunder and slide down into the antlion's waiting jaws. Antlions are the predatory larvae of lacewings.

A number of reptiles live in heaths and pine barrens, including Eastern Box Turtles, nonpoisonous and harmless Eastern

Hognose Snakes, and a toad that also lives in dune areas, Fowler's Toad.

The heaths of the Outer Lands were once home to the Heath Hen, a chickenlike bird closely related to the Greater Prairie Chicken of the midwestern plains. Heath Hens were common at the time the Pilgrims landed, but unfortunately the hens were slow, trusting, and delicious, and they were a popular game animal until the mid-1800s, when they became very rare. The last known group of Heath Hens lived in the heaths and forests of Martha's Vineyard, and all but one male in that tiny band of birds were wiped out in the harsh winter of 1928, and the Heath Hen is now extinct.

Other birds commonly seen in heath areas include Pine and Prairie Warblers, Eastern Bluebirds, Carolina Wrens, Eastern Towhees, Northern Mockingbirds, and the noisy but beautiful Blue Jays. Vesper Sparrows and Grasshopper Sparrows are less common specialists in heaths and grasslands of the Cape and Islands.

Coastal pine barrens
Pine barrens are a unique and uncommon type of forest that develops on dry sandy soils that are often acidic, and poor in

Heath Hen
Tympanuchus cupido cupido

Extinct since 1932

PITCH PINE *Pinus rigida*

PITCH PINE CONES *Pinus rigida*

BEARBERRY *Arctostaphylos uva-ursi*

BEARBERRY, detail

BROOM CROWBERRY *Corema conradii*

BROOM CROWBERRY, detail

ACH HEATHER *Hudsonia tomentosa*

GRAY REINDEER LICHEN *Cladonia rangiferina*

VEET FERN *Comptonia peregrina*

DUSTY MILLER *Artemisia stelleriana*

WBUSH BLUEBERRY *Vaccinium angustifolium*

WINTERGREEN *Gaultheria procumbens*

BEAR OAK *Quercus ilicifolia*

NORTHERN BAYBERRY *Myrica pensylvanica*

BLACK CHERRY *Prunus serotina*

DWARF CHINKAPIN OAK *Quercus prinoides*

BUSHY BEARD LICHEN *Usnea strigosa (inset detail)*

YELLOW WILD INDIGO *Baptisia tinctoria*

HIGHBUSH BLUEBERRY *Vaccinium corymbosum*

SEASIDE GOLDENROD *Solidago sempervirens*

WAVY HAIRGRASS *Deschampsia flexuosa*

SWITCHGRASS *Panicum virgatum*

BLACK HUCKLEBERRY *Gaylussacia baccata*

HYGROSCOPIC EARTHSTAR *Astraeus hygrometricus*

A heath environment: a carpet of Bearberry and Broom Crowberry, with scattered Pitch Pines.

Pitch Pine–Bearberry pine barrens forest type, w Bearberry and Broom Crowberry ground cover.

Pitch Pine–Wavy Hairgrass pine barrens forest type, with Wavy Hairgrass as the dominant ground cover.

Pitch Pine–Blueberry pine barrens forest type, w Lowbush Blueberry as the dominant ground cove.

nutrients. Pine barrens are adapted to frequent wildfires and consist of ground covers and trees that are adapted to withstand and recover from fire damage. Along the East Coast the largest pine barrens are found in central New Jersey, in the Pine Barrens Preserve near Riverhead on eastern Long Island, in the Miles Standish State Forest near Cape Cod, and at Joint Base Cape Cod in Upper Cape Cod. Smaller areas of pine barrens habitat occur in the Cape and Islands area, mostly where coastal heaths are transitioning into coastal Pitch Pine forests along the ocean-facing shores of the Outer Cape at Cape Cod National Seashore.

At the edges of coastal pine barrens there is a continuum from dune or heath habitats into coastal pine barrens habitat, determined by the frequency and height of Pitch Pines in the area. At some point scattered Pitch Pines in the heath or dunes become dense enough to describe the area as a pine forest rather than as dunes or heath. Bear Oak is another essential component of both heaths and pine barrens, but on the poor, dry soils that support these environments, they are mostly shrub-sized in heaths and, in coastal pine barrens, generally an understory shrub or small tree that does not reach the height of the dominant Pitch Pines. Mature coastal pine barrens typically also contain scattered individuals of White Oak, Black Oak, Northern Red Oak, and sometimes Scarlet Oak.

Older pine barrens eventually become mixed Pitch Pines and oak hardwoods forests, as in this upland area of Great Island in Wellfleet. Here the dominant trees are Pitch Pines, but in another 50 years—with no severe fires—this same area could be dominated by mixed oak species and White Pines, with a scattering of Pitch Pines at the margin of the forest. This forest is a reminder that wildfire is an agent in succession. Regular wildfires will preserve this forest as a pine barrens and prevent the oaks from becoming dominant.

Trustom Pond National Wildlife Refuge in coastal western Rhode Island maintains large, grassy fields specifically to encourage and maintain grassland bird, plant, and insect species. In addition to the grasslands, Trustom Pond also contains excellent coastal forest habitat, a large freshwater brackish pond, and a long strand of natural ungroomed beach.

Several types of coastal pine barrens can be seen in the Outer Lands, distinguished by the character of the understory plants beneath the dominant Pitch Pines (see illustrations, p. 254). In areas that were formerly farm fields, the Pitch Pine–Bearberry and Pitch Pine–Wavy Hairgrass forest types predominate, because the soils in former farm fields are poorer in nutrients and often more acidic than average. In lands that were formerly clear-cut woodlots, the Pitch Pine–Blueberry forest type predominates. All three coastal pine barrens forest types are present in the Outer Cape, but the Pitch Pine–Bearberry and Pitch Pine–Wavy Hairgrass types predominate in the Marconi Beach and Marconi Station areas.

Grasslands

Wild grasslands are some of the rarest environments in the Outer Lands today. Native grasslands were never a common type of habitat in the New England landscape before European settlement, but pollen evidence shows that grassland plants were a substantial component of the Outer Lands even before European-style farming and animal husbandry made them the dominant landscape type in eighteenth- and nineteenth-century New England. Before 1620 Native Americans used fire to encourage the growth of grasslands because the interfaces between open grasslands and forest were (and still are) rich in wildlife. These edge habitats between grassland and forest encouraged game like the White-Tailed Deer, an important source of protein for indigenous peoples, particularly in winter. Grasslands were also produced naturally by wildfires, by storm blow-downs of forest areas, and through the activity of North American Beavers. Beavers build dams, of course, and in the process they kill forest trees not only by felling them to create the substance of their dams but also by smothering tree roots in standing water after flooding forest areas. Large rivers in the mainland of New England also produced natural grasslands on floodplains, where periodic flooding discouraged trees. The coastal and island Outer Lands areas have no rivers large enough to support substantial floodplains.

Pipevine Swallowtail
Battus philenor

Further confirmation of grassland habitats in early coastal New England comes from early ornithologists who reported commonly seeing species including Bobolinks, Eastern Meadowlarks, Upland Sandpipers, and Grasshopper Sparrows. Populations of all these grassland birds increased substantially as people clear-cut forests for farming and hay fields.

As farming became less viable in late nineteenth-century New England, farms were abandoned and reverted to forest. In

Eastern Meadowlark
Sturnella magna

Meadowlarks are classic grassland birds now becoming much rarer due to the loss of their primary habitat. A number of Outer Lands coastal parks and Cape Cod National Seashore have made the preservation of remaining grasslands a priority.

Grasshopper Sparrow
Ammodramus savannarum

Upland Sandpiper
Bartramia longicauda

southern New England an untended grassy field will transition to shrubs and then to young forest over 15–20 years. When farming was widespread in southern New England, open grassy hay fields were common and forests were relatively rare, due to the demand for wood for building, heating fuel, and charcoal to fuel small factories.

The largest natural grasslands in the Northeast were the Hempstead Plains of western and central Long Island. This huge natural grasslands, which once covered 40,000 acres, began to vanish in the nineteenth century as farming expanded, but after World War II rapid development of homes, businesses, and other infrastructure destroyed most of the remaining grasslands.

Many New England farms that exist today have moved away from growing hay for animal fodder and now grow more profitable crops such as corn and other vegetables, but even the fodder fields that are left are increasingly unavailable to grassland birds for nesting. Modern farming practices encourage the mowing of hay fields multiple times per year to maximize productivity, yet these midseason mowings often disrupt breeding cycles and destroy grassland bird nests. As an example, current practices to maximize the productivity of alfalfa dictate that the fields be mowed in early to mid-June, which does not leave enough time for grassland-nesting birds such as the Bobolink, Upland Sandpiper, and Eastern Meadowlark to lay their eggs and produce hatchlings that are mobile enough to avoid the mowing blades.

Not just any patch of grass will do for most grassland specialists. Bobolinks require open areas of at least five acres; Savannah and Grasshopper Sparrows won't nest in fields less than 20 acres; and the Upland Sandpiper requires fields that are practically prairies, with 150 acres or more.

Unfortunately, this reversion of the land to forest and the trend away from growing hay has endangered a number of open-land bird species adapted to grasslands. A number of regional conservation efforts are aimed at preserving at least some grassland to support birds and other animals that live only in open field environments. In some parks and nature reserves open grounds are periodically mowed to preserve the grassland and prevent woody shrubs and trees from moving in. Populations of Savannah Sparrows, Bobolinks, Eastern Meadowlarks, Grasshopper Sparrows, and the increasingly rare and endangered Upland Sandpiper have all dropped sharply in recent decades.

Even birds that do not nest in grasslands have become uncommon or endangered through the loss of open fields for feeding and hunting. Northern Harriers, Short-Eared Owls, and Barn Owls have all become less common as grasslands have transitioned to forest, because all three predatory birds feed almost exclusively on Meadow Voles, native mice that live in grasslands and high salt marshes.

Although smaller areas of grassland are scattered throughout the Outer Lands region, the best place to explore grasslands on Cape Cod is in the Fort Hill area of Cape Cod National Seashore, in the town of Eastham. This extensive area of natural grassland was once pasture and later a small golf course,

Top predators like the Barn Owl (*Tyto alba*) are never common, but Barn Owls formerly flourished in the agricultural landscape of 1700s and 1800s New England. Barn Owls need broad, open fields to hunt for mice and voles, and the houses and barns of farms were perfect nesting and roosting places. In today's forested and developed Outer Lands, Barn Owls are increasingly rare.

Barn Owl
Tyto alba

The extensive grasslands of Fort Hill, Cape Cod National Seashore, in Eastham. The blue flower spikes are a garden cultivar variety of Wild Lupine (Lupinis perennis). Fort Hill may be the best location in the Outer Lands to appreciate New England coastal grassland habitat. In the near distance are the Nauset salt marshes and in the far distance the Nauset Spit and Atlantic Ocean. In spring and fall Fort Hill is a superb location to watch migrating birds.

ORGAN PIPE MUD DAUBER *Trypoxylon politum*

TIGER BEETLE *Cicindela sp.*

SEASIDE DRAGONLET *Erythrodiplax berenice*

EASTERN BOX TURTLE *Terrapene carolina*

AMERICAN PAINTED LADY *Vanessa virginiensis*

FOWLER'S TOAD *Bufo woodhousei*

rphotography

AIRIE WARBLER *Setophaga discolor*

dmsphoto

PINE WARBLER *Setophaga pinus*

C Kushner

ROLINA WREN *Thryothorus ludovicianus*

Charles Brutlag

EASTERN TOWHEE *Pipilo erythrophthalmus*

ck Comins

TERN WHIP-POOR-WILL *Antrostomus vociferus*

andromeda108

SPICEBUSH SWALLOWTAIL *Papilio troilus*

but the grasses have been reverting from lawn grasses to wild field grass species since the site became part of Cape Cod National Seashore in the early 1960s.

Fields in winter

Open coastal grasslands, heaths, and salt marshes often attract wintering birds from the Arctic tundra. Northern species such as Snowy Owls, Rough-Legged Hawks, Ross's and Snow Geese, Bohemian Waxwings, Snow Buntings, and Lapland Longspurs all are drawn to snowy fields that resemble their open breeding habitats in northern Canada. In some years, birds like the Snowy Owl (see pp. 192–93) irrupt in unusually large numbers, when abundant food supplies in the Arctic produce more hatchlings, leading many young birds to wander south in winter in search of feeding grounds.

Snow
Bunting

Lapland
Longspur

The classic winter birds of beaches, dunes, heaths, and open grassy areas throughout the Outer Lands. When winter sets in, look for mixed flocks of Lapland Longspurs (*Calcarius lapponicus*), Snow Buntings (*Plectrophenax nivalis*), and Horned Larks (*Eremophila alpestris*).

The Bobolink (*Dolichonyx oryzivorus*) is one of the bird species most threatened by the gradual disappearance of grasslands from New England, especially coastal areas, where the loss of traditional farming and the pressure to develop open land has made this once-common songbird an increasingly unusual sight in the Outer Lands.

Horned
Lark

Painted Turtles bask on a log at Trustom Pond National Wildlife Refuge along the western Rhode Island coast, opposite Block Island.

FRESHWATER

The slow-moving, tidal Pamet River in Truro is more of a long, narrow tidal inlet than a true river. Its eastern headwaters, however, are fresh and surrounded by beautiful riverine shrub habitat.

Visitors to the Outer Lands region are often surprised by its many lakes and ponds, particularly on Cape Cod. It's hardly intuitive that a sandy peninsula stretching 30 miles into the Atlantic would have such an abundant freshwater supply, yet a crucial coupling of chemistry and geology make freshwater not only possible but common. Freshwater is less dense than salt water, and at the scale of large groundwater bodies, freshwater floats above the salt water that permeates the deep layers of glacial sediment under the Cape and Islands (see illustration, p. 69). And although the sandy glacial sediments that make up the Outer Lands are very porous, the huge volumes of sandy earth act like a giant sponge, preventing rainwater and snow meltwater from simply running off the land and into the sea.

Although their origins may be different, most freshwater wetlands on the Cape and Islands are similar in their plant and animal life to ponds and wetlands on the mainland, so this chapter covers only the most distinct aspects of freshwater environments in the Outer Lands and only some of the common species you may encounter there.

Ponds

There is no formal geographic definition of a pond versus a lake, and by tradition even the largest freshwater bodies on Cape Cod are usually called ponds, probably because most of them originated as what geologists call kettle ponds. When the glaciers melted out of the Cape Cod area about 20,000 years ago, they left behind iceberg-sized chucks of ice buried in the glacial till that later melted to form ponds. There are a lot of kettle ponds—depending on how you count them, Cape Cod has more than 500 of all sizes—and from the air the Cape landscape can sometimes appear half liquid with the reflections from these hundreds of ponds. The Cape has many more kettle ponds than the islands because large sections of the southern outwash plains of Martha's Vineyard, Nantucket, and Long Island were never covered by the glacier, and thus there were few blocks of remnant ice to create surface ponds. The Laurentide Ice Sheet covered the entire land surface of what is now Cape Cod, and the glacier left behind thousands of large ice chunks on both the upland moraine areas and the southern outwash plains of the Cape landscape.

Kettle ponds differ from typical ponds in mainland New England in several important ways. As ponds derived from gigantic chunks of ice buried in glacial till, most kettle ponds are unusually deep for their size, with a kettlelike cross section (see illustration, p. 37). The name "kettle" also refers to the rounded shorelines of most kettle ponds. Today most kettle pond shorelines are smooth arcs not because the original glacial ice was rounded but because over thousands of years the natural cycles of wind and water movement have eroded and smoothed the soft earthen pond shores, much as ocean waves and longshore currents smooth ocean beach profiles (see illustration, p. 39).

Silver Spring Pond at Wellfleet Bay Wildlife Sanctuary is unusual in being a spring- and stream-fed freshwater pond rather than a kettle pond. The pond sits in a narrow outwash channel in the Wellfleet outwash plain and was created artificially by a small dam across the streambed.

Ponds in the Outer Lands are also unusual in that they are not usually created or substantially fed by surface streams, as are most mainland ponds. The Cape and Islands land areas are too small and the soils too porous to have upland drainage areas supporting substantial surface streams or rivers. Cape

Cod ponds derive their water almost entirely from groundwater sources, either because the pond is fed by underground springs or because the pond surface intersects the local groundwater level.

Painted Turtles (*Chrysemys picta*) pile onto a log to bask in the sun on Silver Spring Pond at the Massachusetts Audubon Society's Wellfleet Bay Wildlife Sanctuary.

Seasonality in freshwater ponds

Freshwater ponds are not affected by the moon or marine tidal movements, but ponds—especially larger, deeper ponds—are affected by seasonal variations that drive large changes in the water column, moving nutrient-rich, deep cold waters to the surface and drawing surface waters downward to warm and oxygenate the dark pond bottom. Two important physical characteristics of water make pond life possible. Because of its complex crystalline structure, frozen water is less dense than cold water, and ice floats. This counterintuitive fact is crucial, because otherwise ice would coat the bottoms of ponds and it would be much harder for any pond life to survive the winter.

The counterintuitive fact that frozen water is less dense than liquid water is crucial to life in freshwater environments. Ice floats and thereby protects the life underneath it in winter.

Here's how this works. Like most liquids, water becomes denser as it gets colder. But water actually reaches its highest density at 41 degrees Fahrenheit, still well above freezing. As water temperatures fall below 41 degrees, the near-freezing water is less dense than waters above it, and the freezing water rises to the pond surface to form ice. These subtle but important differences in water temperature and density are what drive seasonality in pond circulation, mixing the pond's deeper cold waters even in winter and early spring. As the frozen pond surface warms from 32 degrees to 41 degrees in spring, the denser 41-degree water sinks, displacing the water deeper in the pond.

The combination of warming sun and different water densities drives a process that eventually mixes the whole water column by late spring. In summer deeper ponds stratify, with warm surface waters forming a distinctive layer above cold deep waters. In the fall the temperature layering begins to

Maxar Tamor

Little Cliff Pond is a classic glacial kettle pond in Nickerson State Park, Barnstable.

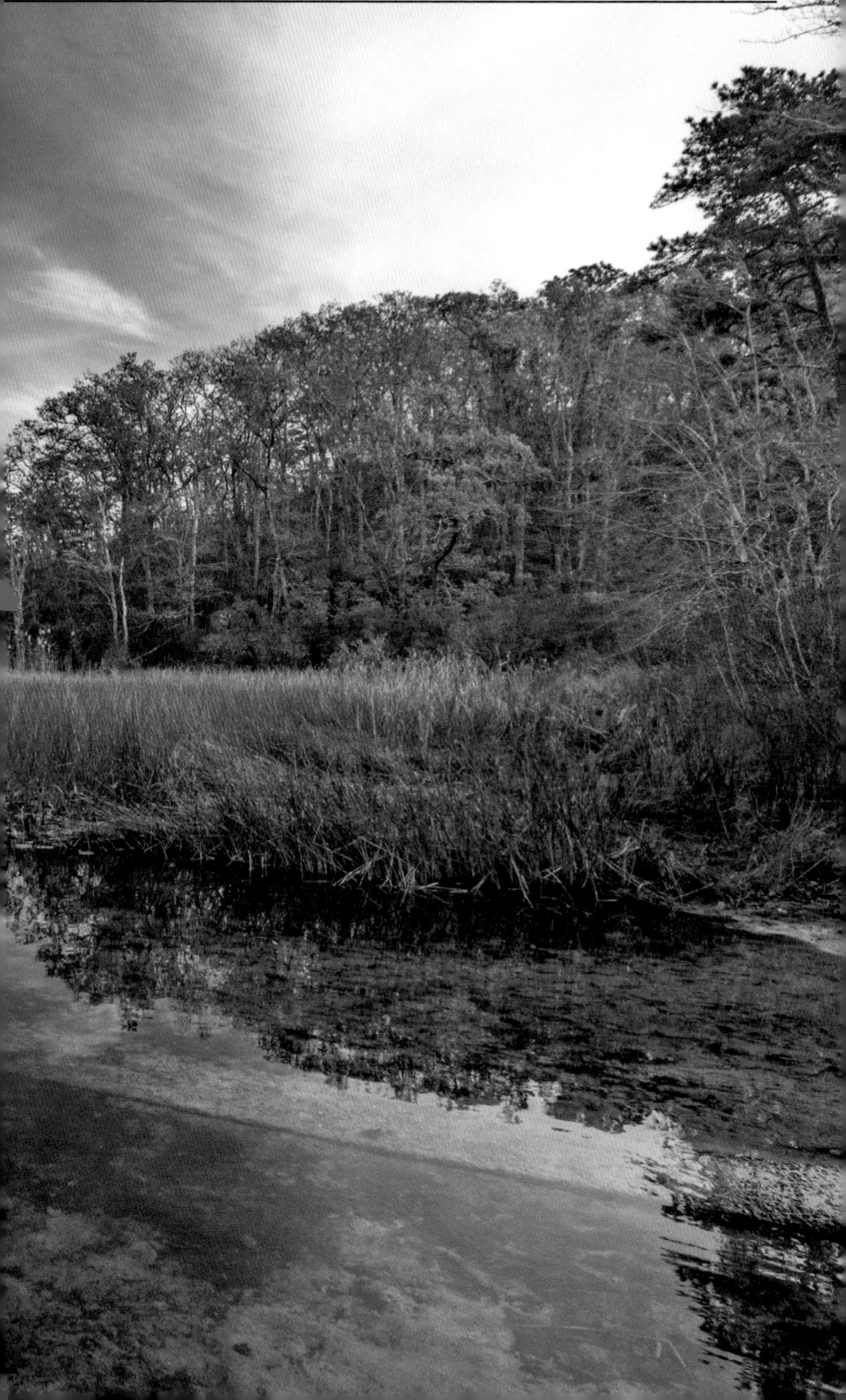

collapse as the surface waters cool and sink, driving a new process that circulates surface and deeper water, preventing stagnation, and bringing both dissolved oxygen and nutrients to all depths of the pond.

Acidity and pollution

The sandy soils of the Outer Lands don't contain the wide range of natural minerals that help buffer mainland aquatic systems, so the water of kettle ponds is often more acid than mainland ponds, shifting the profile of aquatic organisms and plants to those that withstand higher acidity. Many kettle ponds have also been surrounded by Pitch Pines and other conifers for thousands of years, and the conifer needles are yet another source of natural acids. This acidity, combined with a normally scarce nutrient profile, means that Outer Lands ponds are unusually sensitive to pollution and eutrophication caused by excess nitrogen and phosphorus in run-off from sewage systems, home septic fields, and wastewater that contains soap residues. Many Cape Cod ponds turn a deep green in the warmer months because of excess algae growth, and in very polluted ponds there may also be fish kills and deaths of other aquatic animals due to hypoxia—a very low level of dissolved oxygen in the water. As excess nutrients enter the ponds, they cause explosive growth of normally scarce algae, and the algae use up the dissolved oxygen in the water at night, when there is no photosynthesis. The process of eutrophication and hypoxia in ponds is very similar to what happens in coastal marine waters (see discussion, pp. 103–5), but unlike in marine systems there is no daily tidal flushing in ponds, so even a small amount of excess nutrients can do long-term damage.

The Small's Swamp Trail and boardwalk at the Pilgrim Heights location of the Cape Cod National Seashore in Truro is one of the best ways to see a shrub swamp. The short trail descends into a kettle hole depression filled with a small wetland and shrub swamp. In addition to the interesting plant life, the trail makes excellent birding on a sunny morning.

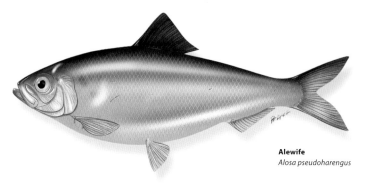

Alewife
Alosa pseudoharengus

Herring runs

Some kettle ponds have an exit stream that carries excess water out of the pond and down to sea level. In precolonial time these pond exit streams played a crucial role in the life cycles of the Cape's anadromous fish—species that breed in freshwater ponds but spend most of their adult lives in the ocean. Each spring so-called river herrings such as the American Shad, Alewife, and Blueback Herring would swim up the exit streams to lay their eggs in the kettle ponds. The Cape Cod streams also once supported the unusual catadromous life cycle of the American Eel, which lives in coastal waters as well as freshwater lakes and streams but breeds deep in the central Atlantic Ocean.

In colonial and early American times these small streams were ideal for building water-powered mills and small factories (see discussion, pp. 83–84), and many Cape Cod streams were dammed, ending the fish runs. Unfortunately, many of those old mill dams are still in place, centuries after the mills rotted away and the dams became useless. In recent years New England states have begun removing or at least breaking open the hundreds of old dams. Restoring the anadromous fish runs is critical not just for preserving Striped Bass, Atlantic Salmon, Alewives, American Eels, and Blueback Herrings. The millions of migrating fish were once an important nutrient source for inland areas near ponds and streams. The fish eggs and the bodies of breeding fish that died or were eaten by birds and other predators along the journey were like a giant nutrient stream, carrying the riches of coastal waters far inland.

The Dexter Grist Mill in Sandwich was built in 1637 and was an operating mill until 1881.

Pond margins and freshwater marshes

Aside from the open waters of ponds, the most common freshwater environments in the Outer Lands are the wetland margins of ponds and the associated freshwater marshes.

Purple Loosestrife
Lythrum salicaria

Most plants growing on pond margins are what biologists call emergent—that is, nonwoody plants that grow in very shallow standing water or permanently wet pond or stream margins. Common Cattail, Arrow Arum, Pickerel Weed, and Purple Loosestrife are all common emergents. Although the showy violet spikes of Purple Loosestrife are pretty, Purple Loosestrife is an aggressive invasive species that displaces native wetland plants, so there are sometimes efforts to pick out the loosestrife plants from wetlands. Another common invasive plant in pond margins is the Common Reed, often called by its generic name, *Phragmites*. *Phragmites australis* is a highly invasive grass from Europe and Asia that has caused problems throughout Northeast wetlands because it displaces native plants that are more valuable to wildlife. Many former cattail marsh areas, for example, are now exclusively *Phragmites* marshes.

In deeper water near pond margins you'll often see floating leaves of two water lily species. The white lilies are Fragrant Water Lily, and the less common yellow lilies are Bullhead Lily. Water lilies root in shallow freshwater, and long stems reach up to anchor round floating leaves and flowers at the water surface.

In ponds near the shoreline, the vegetation may be a mix of the common freshwater pond margin species, along with

more brackish water or salt marsh species such as Marsh Elder (High Tide Bush), Groundsel Tree, and Northern Bayberry (see the chapter "Salt Marshes" for more).

One special type of freshwater wetland area may be seen at the Beech Forest in the Province Lands section of Cape Cod National Seashore. Blackwater Pond and several smaller ponds and associated wetlands are remarkable in that they are mature, well-developed freshwater communities that exist entirely on dune sand, 30 miles from the New England mainland. The sandy Provincelands are a giant sandspit, and there is no bedrock within several hundred feet of the surface of the Beech Forest. The Beech Forest and ponds are a reminder of how the Provincelands looked when the Pilgrims landed in 1620, with dense, mature hardwood and mixed pine-hardwood forests and rich wetland lakes and swamps. In the past the Provincelands were clear-cut and then used to graze livestock, and the forests vanished for hundreds of years. Today's bare sand dunes and sparse Pitch Pine forests indicate an area still in recovery from hundreds of years of human abuse and neglect.

Common Reed
Phragmites australis

Shrub swamps

Shrub swamps evolve in wet bottomland areas that might once have been open ponds but are now filled with vegetation and shallow standing water. Some shrub swamps are

The western end of Blackwater Pond in the Beech Forest of Cape Cod National Seashore, near the Province Lands Visitor Center.

The Atlantic White Cedar Swamp at the Marconi Station area of Cape Cod National Seashore. This fascinating short hike will bring you through one of the rarest wetland environments in the Outer Lands.

seasonally dry, particularly July through September in years with less rain than normal. Small shrub swamp areas exist near most substantial ponds and streams, but Small's Swamp in the Pilgrim Heights area of Cape Cod National Seashore is a particularly good place to explore this freshwater habitat. Small's Swamp sits in a classic kettle depression, or kettle hole. Kettle holes formed the same way kettle ponds did (see earlier in this chapter), but the bottoms of kettle holes do not dip much below the local water table and so do not become open ponds. Shrub swamps are usually dominated by Sweet Pepperbush, Highbush Blueberry, Inkberry, Japanese Honeysuckle, and Black Willow shrubs, with an understory of wetland herbaceous plants such as Skunk Cabbage and ferns including Cinnamon Fern. Common woody vines like Catbrier, Virginia Creeper, and Poison Ivy are usually present, especially in the drier areas surrounding the wet swamp. Red Maples love swampy areas and will invade shrub swamps, eventually becoming dominant and turning the shrub swamp into a Red Maple swamp, as at the excellent Red Maple Swamp Trail at the Fort Hill area of Cape Cod National Seashore.

Skunk Cabbage
Symplocarpus foetidus

Atlantic White Cedar swamps

Atlantic White Cedar swamps are a special type of swamp wetland that was once common but is now rare in the Outer Lands. When the Pilgrims first landed, cedar swamps often occupied kettle depressions that contained shallow standing water for some part of the year and were moist year-round. Most cedar swamps have since been destroyed: Atlantic White Cedar was highly valued for its rot resistance and the swamps were clear-cut of trees. Many former White Cedar swamps are now commercial cranberry bogs.

Fortunately, one of the largest Atlantic White Cedar swamps has been preserved at the Marconi Station area of Cape Cod National Seashore. The trails leading to and from the swamp make a great short hike that leads through various types of Pitch Pine and mixed pine–coastal hardwood forests before descending into the kettle depression of the swamp.

In the swamp the White Cedars form a dense, shady canopy, so the understory is often sparse. The swamp floor is hummocky, with the trees sitting on small mounds surrounded by depressions filled with standing water or peat moss. Scattered Red Maples and Pitch Pines, along with Sweet Pepperbush and Inkberry shrubs, occupy the drier mounds and the periphery of the cedar swamp.

Atlantic White Cedar
Chamaecyparis thyoides

Belted Kingfisher
Megaceryle alcyon

Kingfishers are conspicuous and noisy inhabitants of most wetlands in the Outer Lands, from freshwater ponds and streams to salt marsh creeks. Kingfishers are one of the few birds that can truly hover, a skill they use to good effect hunting above the surface of ponds for small fish.

RiverWalker

Animals of freshwater environments

Insects are not always the top priority when people visit ponds and stream, but if you can look past the biting and stinging kinds of bugs (and wear your bug spray and long pants), insects are one of the most visible and accessible forms of wildlife.

More than 40 species of dragonflies and damselflies are native to Cape and Islands ponds, as are several dozen common butterfly species, and insect-watching is catching on among many birders. On a summer afternoon most wetland birds are roosting out of sight, but the pond edges and marshes will literally be humming with insect life.

Freshwater pond edges and stream banks are the best places to look for the Outer Lands' native frogs and turtles. American Bullfrogs and Green Frogs are both common and look very similar. Green Frogs are indeed usually bright green on the back, with a bright yellow throat and chin. Bullfrogs

Buffleheads are small sea ducks that are common on both freshwater and saltwater ponds and bays. Buffleheads are powerful underwater swimmers that eat insects and plants in freshwater and crustaceans and mollusks in salt water.

are larger and usually have a white or tan throat, along with duller back colors. Both frequent pond edges, and often the only indication of a frog's presence is when one jumps off the banks into the pond as you approach. A good technique is to approach the pond very quietly and study the edges carefully, ideally with binoculars. This will often produce a good view of frogs, turtles, and waterbirds tucked into vegetation on the banks, and it's also a great technique for spotting dragonflies and butterflies that might otherwise scatter if you just blunder noisily onto the pond bank.

Painted Turtles are the Outer Lands' most visible reptiles, since they often bask on logs or stones exposed on the pond surface. Some Painted Turtles bask as much as six hours per day, even on hot summer days. It's sometimes comical to watch as many as a dozen Painted Turtles jockey and jostle for scarce space on a good basking log. The Eastern Box Turtle is also common, but this is a more terrestrial turtle that favors wet forests and muddy areas. The Common Snapping Turtle is in fact common in all kinds of fresh and brackish ponds but is seldom seen except in spring breeding season, when females will often travel overland to lay clutches of eggs just upland of

Great Blue Heron
Ardea herodias

Bufflehead
Bucephala albeola

the pond. Snapping Turtles are major underwater predators in ponds, taking fish, ducklings, other smaller turtles, and even the occasional small mammal. However, Snapping Turtles are quite shy of humans and will almost always swim quietly away from a human bather. The few Snapper bites that get reported each year almost invariably occur when people (well-meaning or not) try to handle Snapper females that are crawling over land to lay eggs. The Common Snapper has a long neck, and simply grabbing the turtle by its shell could result in a nasty bite.

The many freshwater ponds of the Outer Lands are critical for migrating and winter waterfowl of all kinds, but particularly for dabbling ducks, which do not dive under the water surface

Adult female

Adult male

Wood Duck
Aix sponsa

Wood Ducks are common on freshwater ponds and for-ested lake edges throughout the Outer Lands.

to feed and are confined to shallow waters. Mallards, Green-Winged and Blue-Winged Teals, American Wigeons, and Gadwall all feed in the Cape's smaller, shallower ponds. In fall migration many small ponds also attract tiny Pied-Billed Grebes, who feed and rest there before continuing south. Unlike the dabbling ducks, grebes are strong swimmers and divers, and the little Pied-Bills will pop up to the surface suddenly, only to disappear just as quickly in a new dive. Larger ponds attract diving waterfowl in the cold months, including Great and Double-Crested Cormorants and the so-called bay ducks capable of deep dives to feed, including Canvasbacks, Buffleheads, Greater and Lesser Scaup, Common Goldeneyes, and Hooded and Common Mergansers. At any time of year the larger ponds also attract flocks of gulls, which often roost in the central areas of lakes to stay well away from people and predators. If you see a gull flock suddenly lift and scatter over the pond surface, look for a Bald Eagle or Peregrine Falcon nearby.

Pied-Billed Grebe
Podilymbus podiceps

Common Goldeneye
Bucephala clangula

Gadwall
Anas strepera

COMMON CATTAIL *Typha latifolia*

NARROW-LEAVED CATTAIL *Typha angustifo*

PICKEREL WEED *Pontederia cordata*

ARROW ARUM *Peltandra virginica*

BROADLEAF ARROWHEAD *Sagittaria latifolia*

GREATER BLUE FLAG IRIS *Iris versicolor*

EET PEPPERBUSH *Clethra alnifolia*

BUTTONBUSH *Cephalanthus occidentalis*

KBERRY *Ilex glabra*

HIGHBUSH BLUEBERRY *Vaccinium corymbosum*

D MAPLE *Acer rubrum*

BLACK TUPELO *Nyssa sylvatica*

CARDINAL FLOWER *Lobelia cardinalis*

PURPLE LOOSESTRIFE *Lythrum salicaria*

CINNAMON FERN *Osmundastrum cinnamomeum*

JOE-PYE WEED *Eutrochium purpureum*

JEWELWEED *Impatiens capensis*

FRAGRANT WATER LILY *Nymphaea odorata*

MMON WHITETAIL *Plathemis lydia*

GREEN DARNER *Anax junius*

FotoRequest

ONY JEWELWING *Calopteryx maculata*

andromeda108

SPICEBUSH SWALLOWTAIL *Papilio troilus*

erb

ER SWALLOWTAIL *Papilio glaucus*

COMMON BUCKEYE *Junonia coenia*

PAINTED TURTLE *Chrysemys picta*

EASTERN BOX TURTLE *Terrapene carolina car*

COMMON SNAPPING TURTLE *Chelydra serpentina*

GREEN FROG *Rana clamitans*

AMERICAN BULLFROG *Lithobates catesbeianus*

PICKEREL FROG *Lithobates palustris*

SKRAT *Ondatra zibethicus*

MALLARD *Anas platyrhynchos*

AMP **SPARROW** *Melospiza georgiana*

COMMON YELLOWTHROAT *Geothlypis trichas*

)-WINGED BLACKBIRD *Agelaius phoeniceus*

MARSH WREN *Cistothorus palustris*

Common freshwater fish of Outer Lands lakes

To 16 in.

PUMPKINSEED
Lepomis gibbosus

BROOK TROUT
Salvelinus fontinalis

BLUEGILL
Lepomis macrochirus

To 16 in.

YELLOW PERCH
Perca flavescens

To 16 in.

To 28 in.

To 39 in.

Common freshwater fish of Outer Lands lakes

BROWN BULLHEAD
Ameiurus nebulosus

To 21 in.

BROWN TROUT
Salmo trutta

To 40 in.

NORTHERN PIKE
Esox lucius

To 56 in.

SMALLMOUTH BASS
Micropterus dolomieu

To 27 in.

LARGEMOUTH BASS
Micropterus salmoides

To 38 in.

A magnificent oak–pine hardwood forest along Bell's Neck Road in Harwich, part of the Bell's Neck Conservation Lands of brackish and freshwater marshes and mature woodlands.

COASTAL FORESTS

Coastal hardwood forest at Wellfleet Bay Wildlife Sanctuary. Except for an occasional older tree (at right), the woods here are young. Most Cape Cod forests have been clear-cut two or three times since 1620.

The forests of the Outer Lands are young. Almost every forest area in southeastern New England has been clear-cut twice in the past 400 years, and some have been cut three times. Lands immediately along the Atlantic Ocean shores are also the most likely to have been farmed, built upon, paved over, or otherwise disturbed by both human activity and the impact of storms, wildfires, and the sometimes subtle but pervasive damage that salt spray does to most plants near the coast.

The collapse of the Cape and Islands agricultural economy in the nineteenth century dramatically reduced demand for wood, and the abandoned, barren, sandy hills began to reforest in the late 1800s. Today much of the Outer Lands is reforested, albeit with young trees rarely more than 120–130 years old, and most are younger. After a century of reforestation the Outer Lands forests are again under threat, this time from the clearing of forest lots for residential building. The loss is subtle but widespread, as lot by lot, house by house, the forests vanish. Cape Cod alone lost 2,300 acres of forest cover between 2001 and 2011, 70 percent of it to home development. About half that forest loss was in the Upper Cape towns of Bourne, Falmouth, and Mashpee. Although the year-round population of Cape Cod has actually dropped 2 percent in recent years, the average new building lot size is 2.5 acres, as the recent trend has been for ever-larger seasonal homes, at a rate well above national home building trends. This kind of development is what land managers call hard deforestation, because forests lost to homes, commercial development, and roads are gone forever.

On the Outer Cape things look better for wild forests, if only because so much of the Outer Cape is preserved within the Cape Cod National Seashore. Ironically, the National Seashore land managers have sometimes struggled with the paradox of wild grasslands and heaths being invaded by forest trees, converting those open lands to coastal forest, which decreases the ecological diversity of the Outer Cape.

Outer Lands forests
Most forests in the Outer Lands are of two types: Pitch Pine–dominated coastal pine barrens on dry, nutrient-poor lands, and Pitch Pine–mixed oak forests on upland or inland areas away from salt spray.

Rose-Breasted Grosbeaks
Pheucticus ludovicianus

The Outer Lands forests are rich with bird life in all seasons, but particularly in spring. Male and female grosbeaks on Canadian Serviceberry (Shadbush, *Amelanchier canadensis*).

These coastal forests have three distinct elements:

• A transitional edge dominated by low shrubs, brambles, and trees stunted by salt spray or very dry conditions.

• True forest areas dominated by taller trees and with a forest understory or ground cover, still under the influence of some salt spray.

• Inland forest, away from salt spray, often on moister grounds of former woodlots, not nutrient-exhausted former farms.

A small grove of Sassafras trees (*Sassafras albidum*) showing their characteristic twisted trunks. Sassafras is one of the most common hardwoods in coastal forests.

Transitional beach, dune, and marsh edges

The upper marsh and beach border is often a prickly tangle of shrubs, vines, and pioneer plants that specialize in transitional areas and disturbed ground. Marsh Elder (High Tide Bush), Groundsel Tree, Northern Bayberry, Black Cherry, Eastern Redcedar Juniper, and Shining (Winged) Sumac are common native shrubs in areas that are near salt water and regularly receive salt spray and occasional soakings with salt water in major storms.

In more sheltered border areas, Staghorn Sumac and Smooth Sumac join the mix of shrubs, and particularly in areas recently disturbed by human activity, the invasive sumac look-alike Ailanthus (Tree-of-Heaven) may be abundant. Aside from Black Cherries that manage to grow beyond bush height, the most prevalent small trees are Eastern Redcedar Junipers, a species that is common in all Outer Lands coastal environments from salt marshes to coastal bluffs and sand-

(Continued on p. 308)

A young Pitch Pine forest on Great Island, Wellfleet. As recently as the early 1900s this area was open pastures and farm fields. The dominant Wavy Hairgrass ground cover is an indication that the area was farmed and that the farming exhausted the soil's nutrients.

POISON IVY *Toxicodendron radicans*

VIRGINIA CREEPER *Parthenocissus quinquefolia*

BEAR OAK *Quercus ilicifolia*

CATBRIER *Smilax glauca*

FOX GRAPE *Vitis labrusca*

JEWELWEED *Impatiens capensis*

WBERRY *Rubus sp.*

WINEBERRY *Rubrus phoenicolasius*

IATIC BITTERSWEET *Celastrus orbiculatus*

BLACK SWALLOW-WORT *Cynanchum louiseae*

PANESE HONEYSUCKLE *Lonicera japonica*

AILANTHUS (TREE-OF-HEAVEN) *Ailanthus altissima*

JAPANESE KNOTWEED *Fallopia japonica*

COMMON REED *Phragmites australis*

AUTUMN OLIVE *Elaeagnus umbellata*

MULTIFLORA ROSE *Rosa multiflora*

FIELD BINDWEED *Convolvulus arvensis*

BLACK LOCUST *Robinia pseudoacacia*

MARSH ELDER *Iva frutescens*

GROUNDSEL TREE *Baccharis halimifolia*

SWITCHGRASS *Panicum virgatum*

NORTHERN BAYBERRY *Myrica pensylvanica*

HIGHBUSH BLUEBERRY *Vaccinium corymbosum*

EASTERN REDCEDAR JUNIPER *Juniperus virginiana*

BEACH PLUM *Prunus maritima*

BLACK CHERRY *Prunus serotina*

QUAKING ASPEN *Populus tremuloides*

STAGHORN SUMAC *Rhus typhina*

SHINING (WINGED) SUMAC *Rhus copallina*

BEAR OAK *Quercus ilicifolia*

SAFRAS *Sassafras albidum*

BLACK OAK *Quercus velutina*

RTHERN RED OAK *Quercus rubra*

WHITE OAK *Quercus alba*

ERICAN HOLLY *Ilex opaca*

RED MAPLE *Acer rubrum*

CANADIAN SERVICEBERRY *Amelanchier canadensis*

COMMON APPLE *Malus pumila*

UMBRELLA SEDGE *Cyperus strigosus*

AMERICAN LINDEN (BASSWOOD) *Tilia amer*

ARROWWOOD VIBURNUM *Viburnum dentatum*

BLACK TUPELO *Nyssa sylvatica*

WERING DOGWOOD *Cornus florida*

INKBERRY *Ilex glabra*

PLELEAF VIBURNUM *Viburnum acerifolium*

SWAMP AZALEA *Rhododendron viscosum*

MMON MILKWEED *Asclepias syriaca*

WAVY HAIRGRASS *Deschampsia flexuosa*

YARROW *Achillea millefolium*

ORCHARD GRASS *Dactylis glomerata*

ORCHARD GRASS, flowers

FOXTAIL GRASS *Alopecurus sp.*

WILD GERANIUM *Geranium maculatum*

PINK AZALEA *Rhododendron periclymenoide*

LDENROD *Solidago sp.*

RUGOSA ROSE *Rosa rugosa*

TURE ROSE *Rosa carolina*

SWAMP ROSE MALLOW *Hibiscus moscheutos*

ITE WOOD ASTER *Eurybia divaricata*

LATE PURPLE ASTER *Symphyotrichum patens*

Great Horned Owl
Bubo virginianus

The Great Horned Owl is a top predator in Outer Lands forests and probably the most common large bird of prey in the area besides the Osprey and Red-Tailed Hawk. These owls are nocturnal hunters, mostly of small mammals such as squirrels and skunks, but they also sometimes take ducks and other medium-sized birds.

spits. Junipers are remarkably tough, but Hurricanes Irene (2011) and Sandy (2012) battered many marsh and shoreline junipers on the Outer Lands coasts. These junipers still show the effects of severe salt spray damage, where the windward side of the tree (usually the side that faces the ocean, or the southeast) has a lot of dead or salt-burned foliage. Pitch Pines near ocean coasts also commonly show salt-killed red-brown needles and dead branches.

The shrubs and stunted trees of the beach and marsh margins support a dense tangle of bramble species, vines, and small softwood trees like sumacs. Poison Ivy, Virginia Creeper, Fox

Grape, Dewberry (*Rubus sp.,* often collectively called wild raspberries), Catbrier, and the very similar Bullbrier can form bramble hedgerows so thick and thorny that few animals can pass through them. Unfortunately, much of the Outer Lands coastal area has been disturbed by human activity many times over the past several centuries. Farming, road building, dredge spoil dumping, salt marsh filling, and other activities destroy natural plant communities and allow invasive nonnative species to move in and dominate the disturbed ground. Asiatic Bittersweet, Multiflora Rose, Japanese Honeysuckle, Autumn Olive, Wineberry, Ailanthus, and Japanese Knotweed are all nonnative plants that are commonly seen in salt marsh, beach, and coastal forest border areas.

Coastal forests

The coastal forests of the Outer Lands are a distinct assemblage of tree and understory species, where such trees as Sassafras, Canadian Serviceberry (Shadbush), American Holly, American Linden (Basswood), Quaking Aspen, Red Maple, and other species are more common than you would see in inland forests. Many of the coastal forest tree species are what ecologists call early successional or pioneer species, plants that specialize in moving into disturbed grounds or marginal habitats. These fast-growing, relatively small trees are gradually replaced in mature pine-oak-hickory-maple hardwood forests but tend to persist in areas near the coast. The presence of these pioneer trees also reminds us that the coasts of southeastern New England have been heavily modified by human activity and that even the best-protected coastal forests are relatively young. A century ago most of what is now coastal forest was open farmland or logged-out woodlots that were gradually abandoned as coal and oil replaced wood for winter heating and better transportation systems made midwestern farms far more competitive than New England and Long Island farms.

Upland forest of mixed oaks, Sassafras, Red Maples, and Pitch Pines at Nickerson State Park in Brewster.

One of the most common broadleaf trees in coastal forests is Sassafras, with its distinctively twisted trunks and (mostly) mitten-shaped leaves. Young Sassafras trees often line coastal woodland hiking trails. In the spring, the many Canadian Serviceberry trees (locally called Shadbush because they blossom in spring when shad ran in local streams) are white with blooms, joined by a few Flowering Dogwoods, as well as by Common Apples that survive from abandoned coastal fruit orchards. In more mature coastal forests White, Black, and Northern Red Oaks and mature Red Maples are the largest trees. Often these large maples and oaks predate the rest of the coastal forest and were once field trees growing along farm walls and roads. The remains of old stone walls that formerly

lukicarbol

Long-Eared Owl
Asio otus

marked farm field borders now run through forests that grew up when the farms were abandoned in the late 1800s.

The understory growth at the edges of coastal forests is a combination of marsh and beach edge species plus common coastal thicket plants. At marsh and beach edges Marsh Elder (High Tide Bush), Groundsel Tree, Northern Bayberry, and Black Cherry shrubs and small trees predominate. The invasive shrub Autumn Olive and the invasive vines Black Swallow-Wort and Japanese Honeysuckle are sadly common, driving out native plants that have more food value for small animals. All three common native sumacs, Shining (Winged), Smooth, and Staghorn, occur along the paths that receive some direct sunlight and produce valuable fruit and seeds for migrants. In open areas along the paths and in small clearings Common Milkweed and Deer-Tongue Grass are common, although fields shift in plant composition from year to year. For example, Jewelweed may dominate the open areas in some years but will fade back to a secondary role in others. Switchgrass, Wavy Hairgrass, and Foxtail Grass are common grasses, as are Umbrella Sedges, with their distinctive flowers that look like short bottlebrushes.

Transition to mature inland oak-pine forest

The region's coastal forests share almost all the same tree species as inland forests, but with a much higher percentage of smaller pioneer species. The transition from coastal forest to inland forest, therefore, can be subtle, especially if you look only at the trees. Often the most noticeable transitions into inland forest are in the understory. Classic eastern forest understory species such as Spicebush and Mapleleaf Viburnum are uncommon in coastal forests and become dominant forest understory species only when well away from salt spray or saline soils. Most ferns are similarly uncommon in coastal forests, growing into lush ground cover only in upland forests removed from salt spray.

Animals

Many of the birds and mammals that feed in the upper marsh shelter and nest in the adjacent coastal forests. Great Horned Owls, Long-Eared Owls, and the tiny Northern Saw-Whet Owl all hunt in salt marshes, open fields, woodlands, and forest margins but roost in nearby coastal forests, as does that versatile predator of all coastal habitats, the Black-Crowned Night-Heron.

Spring migrant birds

In spring coastal woods attract a variety of migrating woodland birds. Beginning in March, Red-Winged Blackbirds, Common Grackles, Northern Cardinals, and Marsh Wrens

Brown Creeper
Certhia americana

announce the coming of warm weather. Noisy gangs of boreal songbirds heading north to the Canadian woods move through the trees, including Golden-Crowned and Ruby-Crowned Kinglets, Brown Creepers, Black-Capped Chickadees, and White-Throated Sparrows.

On a moist, early May morning, with warm-front winds from the southwest, large groups of songbirds will move through the woodlands, including Red-Eyed Vireos, Veerys, Yellow Warblers, Yellow-Rumped Warblers, American Redstarts, and Common Yellowthroats. Carolina Wrens, Marsh Wrens, and Song Sparrows travel through as migrants but also remain to breed in Outer Lands coastal woodlands.

Spring is best for viewing migrants on the Cape because of the Cape Cod Effect. As a peninsula 30 miles out at sea, the Cape is a blind alley for northbound migrating birds. The northbound birds follow the southern New England coastline and funnel northeast through the Mid- and Outer Cape until they hit Provincetown, where they discover that they are surrounded by water. The birds drop down into areas such as the Beech Forest and mill about, sometimes for days, before braving the flight across Cape Cod Bay to the mainland.

Fall migrants

In autumn many of the same woodland species move south again, but fall migration also brings large flocks of Blue Jays, Tree Swallows, and all of the common blackbird species.

American Kestrel
Falco sparverius

Male

Female

**GOLDEN-CROWNED
KINGLET**
Regulus satrapa

L 4 in.
WS 7 in.

**RUBY-CROWNED
KINGLET**
Regulus calendula

L 4.25 in.
WS 7.5 in.

Common in migration,
particularly in autumn

**RED-BREASTED
NUTHATCH**
Sitta canadensis

**WHITE-BREASTED
NUTHATCH**
Sitta carolinensis

L 5.75 in.
WS 10.5 in.

L 4.5 in.
WS 8.5 in.

L 5.25 in
WS 8 in.

**BLACK-CAPPED
CHICKADEE**
Poecile atricapillus

L 11 in.
WS 16 in.

BLUE JAY
Cyanocitta cristata

WHITE-THROATED SPARROW
Zonotrichia albicollis

L 6.75 in. WS 9 in.

Common in migration, particularly in autumn

L 6.25 in.
WS 8.25 in.

SONG SPARROW
Melospiza melodia

WHITE-CROWNED SPARROW
Zonotrichia leucophrys

L 7 in. WS 9.5 in.

GRAY CATBIRD
Dumetella carolinensis

L 8.5 in. WS 11 in.

CAROLINA WREN
Thryothorus ludovicianus

L 5.5 in.
WS 7.5 in.

MARSH WREN
Cistothorus palustris

L 5 in.
WS 6 in.

HOUSE WREN
Troglodytes aedon

L 4.75 in.
WS 6 in.

Male

Female

NORTHERN CARDINAL
Cardinalis cardinalis

L 8.75 in. WS 12 in.

NORTHERN MOCKINGBIRD
Mimus polyglottos

L 10 in. WS 14 in.

EASTERN SCREECH OWL
Megascops asio

L 8.5 in. WS 20 in.

SHARP-SHINNED HAWK
Accipiter striatus

Common in migration, particularly in autumn

L 11 in. WS 23 in.

Cooper's Hawk
Accipiter cooperii

Hermit Thrush
Catharus guttatus

Yellow-Rumped Warbler
Setophaga coronata

American Redstart
Setophaga ruticilla

Yellow Warbler
Setophaga petechia

Often the flocks flow over the coasts, heading southwest in a continuous stream across the sky. But if the flocks do drop into the trees, the experience can be amazing, as 300–400 noisy Blue Jays suddenly blast a riotous mix of jay calls and blue-and-white blurs across the treetops.

All the songbirds in the trees draw down migrating hawks, particularly forest bird hunters such as the common Sharp-Shinned Hawk. Watch for these sleek, short-winged pursuers as they fast-cruise through the forest canopy, looking for an unwary songbird to pick off.

In autumn the woodlands like the Beech Forest in Provincetown can be full of blackbirds, wood warblers, and Cedar Waxwing flocks. Palm and Yellow-Rumped Warblers are common in both spring and fall migration. The thickets along the wood edges draw Gray Catbirds, Northern Mockingbirds, and a range of sparrow species. Later in the fall Ruby-Crowned and Golden-Crowned Kinglets move through park woods, and some linger well into winter, joining gangs of Black-Capped Chickadees, Tufted Titmice, and Brown Creepers in winter foraging flocks.

In winter it's worthwhile to scan clusters of Eastern Redcedar Junipers and other dense conifers along the edges of the woods near salt marshes or open fields for roosting Northern Saw-Whet Owls.

Spring and fall hawk-watching

The Cape and Islands are not as popular as mainland New England sites for fall hawk-watching, but that doesn't mean the hawks aren't there. Open coastal areas such as Fort Hill in Eastham, Sandy Neck in Barnstable, Marconi Station, Salt Pond Visitor Center in Eastham, Province Lands Visitor Center in Provincetown, and, most famously in the spring, the Beech Forest area of Cape Cod National Seashore in Provincetown all provide excellent viewing for spring or fall migration.

As mentioned above, the Cape Cod Effect also funnels northbound hawks into the Outer Cape, and hawks share the same reluctance songbirds have for crossing long stretches of water, so the forests of the Outer Cape in places like Wellfleet Bay Wildlife Sanctuary and the Beech Forest in Provincetown can produce excellent viewing.

Throughout New England the best spring migration days occur when warm fronts move up from the south. The moist, warm fronts often bring overcast, drizzly, or hazy weather, but they also sweep birds up the Atlantic Coast, riding the favorable tailwind northward. In fall the opposite is true: cold

Rusty red breast; wing linings more white than in the Red-Shouldered

BROAD-WINGED HAWK
Buteo platypterus

L 15 in. WS 34 in.

Broad white bands in the tail

Large, evenly spaced dark-light bands on tail

RED-SHOULDERED HAWK
Buteo lineatus

Thin white bands on a dark tail

L 17 in. WS 40 in.

Tail dark with thin white bands

Rusty red breast and underwings

Dark belly band

RED-TAILED HAWK
Buteo jamaicensis

Brick-red tail

L 19 in. WS 49 in.

Brick-red tail in the adult; note belly band

Red-Eyed Vireo
Vireo olivaceus

American Goldfinch
Spinus tristis

Pine Warbler
Setophaga pinus

Tufted Titmouse
Poecile atricapillus

Blue Jay
Cyanocitta cristata

fronts moving in with clear, cool days and winds from the northwest sweep southbound birds into coastal New England, where they then follow the shoreline west and south toward Rhode Island, Connecticut, and Long Island. On such days the south shore of Long Island can also produce excellent birding, as tens of thousands of birds funnel along the barrier islands rather than brave the long crossing of the New York Bight to get to New Jersey and points south.

As the sun heats the landscape, the warm ground heats the air near the ground, and that warm air forms columns of rising air called thermals. Most migrating hawks circle in these thermals to gain altitude and then glide relatively effortlessly on their way, instead of having to use active flapping flight to travel the thousands of miles from northern North America to the southern United States, Central America, and South America. Warm thermals form over solid ground but not over the relatively cold waters of Cape Cod Bay or the Atlantic Ocean. Hawks are understandably reluctant to fly out over water, where they get no assist from rising thermals and must actively flap across the water to the next piece of land.

Start in the early morning to get the best views of migrating songbirds and hawks moving low over marshes and forests. The early hawks will often be looking for a breakfast of songbirds, so a morning walk through coastal woodlands will often be rewarded with views of Sharp-Shinned and Cooper's Hawks rocketing through the woods below treetop level, looking for a meal. Later in the day the movement of soaring groups of hawks becomes harder to view, because the thermals rising above the coastline bring the hawks up so high that they become difficult to spot even with binoculars.

Resident birds

The breeding coastal woodland birds are the common but beautiful residents of most woodlands and forest edges of the Atlantic Coast. Song Sparrows, American Goldfinches, Baltimore Orioles, and Northern Cardinals inhabit forest edges. Birding in bramble areas will often yield White-Throated Sparrows, Eastern Towhees, and the tiny but gorgeous Common Yellowthroat. Overhead in the crown foliage of the trees, look and listen for Red-Eyed Vireos. These vireos are the most common bird species in the deep woodlands during the summer, but they are heard more often than seen because they favor the treetops for feeding and nesting. Watch the trees for flashes of crimson from American Redstart warblers, common but not easy to spot high in the trees. Forest edges and marshes ring with the calls of Red-Winged Blackbirds from mid-March until they depart southward in late fall.

The Beech Forest in the Province Lands section of Cape Cod National Seashore is one of the best birdwatching areas on Cape Cod, particularly during spring and fall migration.

CEDAR WAXWING Bombycilla cedrorum

RED-BELLIED WOODPECKER Melanerpes carol

brm1949

COMMON YELLOWTHROAT Geothlypis trichas

Paul Reeves Photography

AMERICAN ROBIN Turdus migratorius

Rachelle Vance

DOWNY WOODPECKER Picoides pubescens

Gregg Williams

MOURNING DOVE Zenaida macroura

Mammals

Coastal forests are usually home to the same range of forest mammals as more inland forests. Most forest and coastal edge mammal species are nocturnal or active only at dawn and dusk, so you won't often see such common residents as White-Tailed Deer, Raccoons, Virginia Opossums, or Northern Flying Squirrels. The structural height complexity of woodland edges (dense foliage from ground level all the way up to the treetops), the diverse range of plant species, and the shelter that dense bramble provides all make coastal woodlands particularly attractive to small mammals—and to the animals that hunt them. White-Footed Deer Mice, Eastern Chipmunks, Raccoons, Gray Squirrels, American Red Squirrels, and Northern Flying Squirrels are all common. Red Foxes were once more common in Outer Lands coastal forests, but with the rise of the Eastern Coyote population in the past 20 years, foxes are now a bit less common. White-Tailed Deer thrive in edge habitats with a rich supply of shrub-height plants and thus are common in coastal woods, though they are not often seen by casual hikers. Eastern Cottontail rabbits and Groundhogs (Woodchucks) are very common in coastal woodland edges, in open meadows, and along grassy roadsides near bramble thickets, where they can quickly retreat from predators.

Both of the region's vulture species can be seen over most of the Outer Lands, but only the Turkey Vulture is common over Cape Cod. Both the Turkey Vulture and the smaller Black Vulture are formerly more southern birds that have moved steadily up the Atlantic Seaboard as the climate has warmed over the past 50 years. On Long Island and in Rhode Island Black Vultures are common. The Black Vulture is the newer arrival, but in the Cape and Islands the Black Vulture is still unusual. However, that is likely to change as the climate continues to warm, and every year yields more Black Vulture sightings on the Cape.

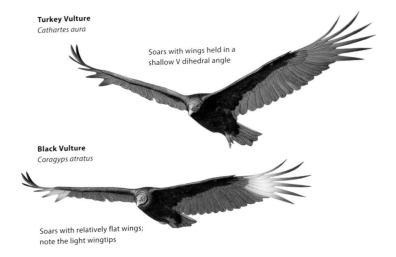

Turkey Vulture
Cathartes aura

Soars with wings held in a shallow V dihedral angle

Black Vulture
Coragyps atratus

Soars with relatively flat wings; note the light wingtips

Nicolase Lowe

WHITE-TAILED DEER *Odocoileus virginianus*

hkuchera

RACCOON *Procyon lotor*

DMM Photography Art

WHITE-FOOTED DEER MOUSE *Peromyscus leucopus*

elharo

EASTERN CHIPMUNK *Tamias striatus*

Cihan Ciin

GRAY SQUIRREL *Sciurus carolinensis*

Anterovium

AMERICAN RED SQUIRREL *Tamiasciurus hudsoni*

RTHERN FLYING SQUIRREL *Glaucomys sabrinus*

RED FOX *Vulpes vulpes*

STERN COYOTE *Canis latrans var.*

EASTERN COTTONTAIL *Sylvilagus floridanus*

OUNDHOG (WOODCHUCK) *Marmota monax*

LONG-TAILED WEASEL *Mustela frenata*

Coast Guard Beach, Truro.

SHALLOW COASTAL WATERS

Striped Bass (*Morone saxatilis*)

Below the low tide lines of the Outer Lands there are four main aquatic habitats. The subtidal zone and shallow bottom areas, and the unique and important habitat formed by Eelgrass beds, are considered here; the deeper coastal waters are described in the next chapter.

The subtidal zone

The subtidal zone lies from the lowest tide line down to about 20 feet. At this level life is truly aquatic, although many crabs visit lower intertidal zones in search of food, particularly at night. Because these permanently submerged shallows receive enough sunlight for plant growth, most seaweeds (macroalgae) live in this zone, as do Eelgrass and Widgeon Grass. Wave action shapes this environment both physically, through the constantly churning waves and their resulting shore currents, and chemically, through supplying oxygen and nutrients. For sessile (fixed in place) filter feeders like barnacles and mussels, the constant water movement is critical to life, ensuring a steady supply of food. Beyond 10 feet of depth, the algae and sea grass populations drop sharply owing to the low light levels. At about 20 feet, the properties of wave action, superheating in summer, supercooling in winter, and intertidal species' depth limits quickly make the environment more like deeper bottom conditions.

Border zones between one environment and another, or ecotones, are the most productive environments. As a transition zone between tidal areas and deeper waters, the subtidal zone is one of the most productive zones of the ocean. Many open-water fish spend early life in the relative safety of shal-

Many wading birds such as the Great Egret (*Ardea alba*) frequent shallow marsh creeks, tidal flats, and subtidal areas hunting for small fish, crabs, and other small invertebrates.

lows, particularly where Eelgrass beds and rocky bottoms with crevices offer hiding places. Larger predators such as Bluefish and Striped Bass sweep through the shallows for prey, and many wading and diving birds specialize in picking off unwary crabs, fish, and shrimp that live in a few feet of water. Ospreys glide high above, looking for the slightest movements from such favored prey as small flounders and Atlantic Menhaden.

The subtidal zone bottom is rocky with glacial boulders and gravel in some areas, sandy or muddy in others. Although many subtidal creatures frequent both areas, most plants and animals specialize in either soft sediments or rocky bottoms.

Plants

Green plants need sunlight to perform photosynthesis, and light doesn't penetrate far into water, especially when the water is naturally cloudy with phytoplankton and silt from beaches and tidal flats. The subtidal zone is thus home to larger marine plants. Most of these are macroalgae (seaweeds) that attach to the bottom, in contrast to single-celled algae, which float freely in the ocean and bay waters.

Seaweeds come in three basic varieties, loosely grouped by their dominant color: green, red, or brown. The most common green species, Sea Lettuce, is the filmy, bright green algae that is ubiquitous on beaches, tidal flats, and at salt marsh edges. Another common green seaweed is Green Fleece, seen on submerged glacial boulders, on sandy bottoms, and in salt marshes. Green Fleece, which is not native to the United States, spread from the eastern shores of Asia to Europe by attaching to the hulls of sailing ships and reached North America in 1957. The bright green algae that commonly covers dock pilings and boulders in both the lower intertidal and subtidal zones is called Gut Weed. Stone Hair is a finer-grained, shorter-stranded algae that also forms bright green mats on intertidal and subtidal rocks, breakwaters, and pier

Bluefish
Pomatomus saltatrix

Steve Bower

pilings, often in association with the more coarse-grained Gut Weed. Be wary of walking on exposed boulders covered in algae; even when the surface looks dry, it can be slippery underneath.

Brown algae are often the most visible and familiar marine plants in rocky areas. Rockweed and Knotted Wrack are the two most common species, seen on seaside boulders, docks, seawalls, or any other fixed structures along the shore. Rockweed and Knotted Wrack also grow in salt marsh creeks where they can find a secure anchor to grow from. Both species grow from the low intertidal zone well into the subtidal zone, down to a depth of about three feet. Sugar Kelp and Atlantic Kelp grow in the deeper areas of the subtidal zone, from five to 20 feet or more, using their long stems to keep their long, flat blades in the well-lit waters near the surface. Kelp beds are not usually visible from shore, but kelp fronds commonly wash up on beaches. Irish Moss is a common red algae that normally sits right at and just below the lowest tide line, in areas where glacial boulders or other hard structures give it a solid base to grow from.

Subtidal invertebrates and fish
In marine biology, the collection of plants and animals that live on, under, or near the bottom of a body of water is called the benthic community. Subtidal areas with sandy or muddy bottoms have a rich infauna of animals that burrow into the bottom sediments for shelter. Most clams and marine worms, some crabs, and even some fish bury themselves at least

The Osprey (*Pandion haliaetus*) is one of the most visible birds that feed in the subtidal zone. Almost extirpated in the northeastern coastal region owing to DDT contamination, the Osprey has made a remarkable comeback in the past 30 years, and populations are nearing total recovery.

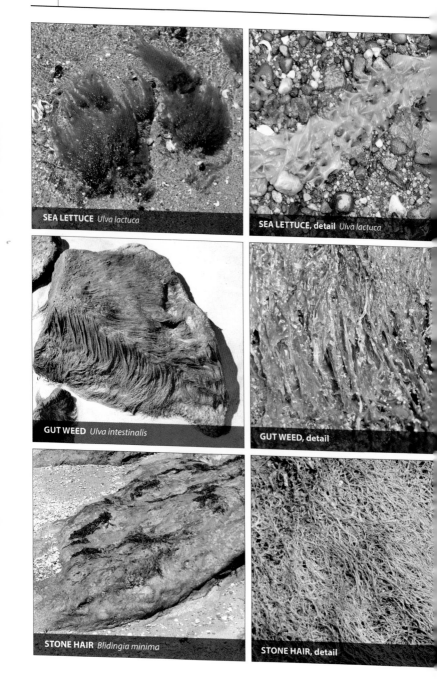

SEA LETTUCE *Ulva lactuca*

SEA LETTUCE, detail *Ulva lactuca*

GUT WEED *Ulva intestinalis*

GUT WEED, detail

STONE HAIR *Blidingia minima*

STONE HAIR, detail

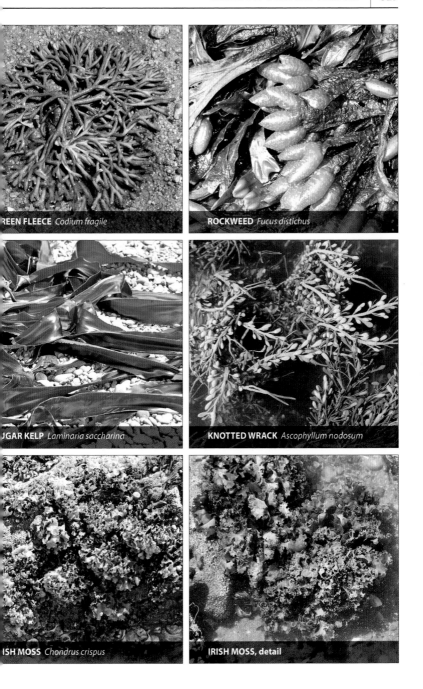

REEN FLEECE *Codium fragile*

ROCKWEED *Fucus distichus*

JGAR KELP *Laminaria saccharina*

KNOTTED WRACK *Ascophyllum nodosum*

ISH MOSS *Chondrus crispus*

IRISH MOSS, detail

Northern Quahog
Mercenaria mercenaria

Invertebrates in and near the bottom sediments of the subtidal zone.

partially as protection in flat bottom areas that often lack rock or plant shelter. Animals that live primarily on or just above the bottom surface are called epifauna. These include most of the familiar shoreline crabs, which welcome rock crevices or Eelgrass patches as shelter but don't normally dig burrows in the warmer months.

Clams

Soft sediment bottoms and tidal flats such as those that line the southern shore of Cape Cod Bay can look deceptively lifeless unless you look for the single or paired siphons of clams buried within them. Northern Quahogs have short, paired siphons, and the top edge of their shells is rarely buried more than an inch below the surface. In smooth sand or mud bottoms look for a figure eight of the twin open siphon holes. Quahogs are called by a variety of names based on their size, but little necks, cherrystones, chowder clams, and quahogs are all the same species: the Northern Quahog. Softshell Clams, or steamers, have an extremely long, tough pair of siphons encased in a thick black membrane. The long siphons allow the Softshell Clams to bury themselves far below other clams, sometimes 10 inches deep. The Atlantic Jackknife Clam, or razor clam, has a very short siphon that looks keyhole-shaped at the surface. These clams bury themselves vertically

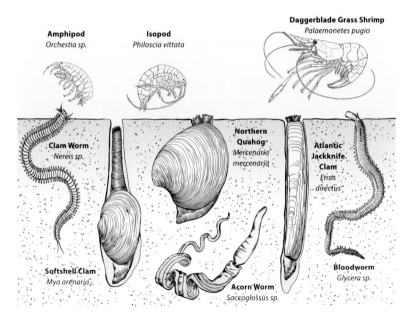

Daggerblade Grass Shrimp
Palaemonetes pugio

Amphipod
Orchestia sp.

Isopod
Philoscia vittata

Clam Worm
Nereis sp.

Northern Quahog
Mercenaria mercenaria

Atlantic Jackknife Clam
Ensis directus

Softshell Clam
Mya arenaria

Acorn Worm
Saccoglossus sp.

Bloodworm
Glycera sp.

with a short but strong foot on the lower end opposite the siphon. Atlantic Jackknife Clams sometimes pop up above the surface, often when disturbed by mud worms probing their burrows from below. If a Jackknife Clam senses movement, it disappears in a flash down into its burrow.

The largest bivalve in these waters, the Atlantic Surf Clam, lives in sandy bottoms from the subtidal zone down to the deep offshore waters. This clam is surprisingly long-lived, living for 31 or more years. Most Surf Clams are harvested at about 15–20 years, primarily for chowders and fried clams. Their chief predators are Moon Snails, Horseshoe Crabs, Atlantic Cod, and, of course, humans.

Softshell Clam
Mya arenaria

One of the most common shelled animals of the subtidal zone is a sea snail, the Slipper Shell, often called a boat shell. These snails are common in both soft sand and rocky shallow subtidal areas and on flats and beaches exposed at low tide, especially in more sheltered bays and inlets. They are filter feeders and typically live in stacks, with older individuals at the bottom and successive layers of younger individuals attaching on top of the older shells. A strong, muscular foot holds each individual in place in the stack, and when submerged, the foot relaxes slightly to open a gap through which the snail draws water to filter for plankton.

Whelks

Whelks are nocturnal sea snails that prey on clams, oysters, and other bivalves. Channeled Whelks prefer sandy, shallow subtidal areas, where they can be common. The similar but typically larger Knobbed Whelk favors deeper waters. In summer whelks avoid warm waters by moving to deeper waters, but in spring and early summer, some Knobbed Whelks migrate into the subtidal zone to feed.

Atlantic Jackknife Clam
Ensis directus

Eastern Oyster

The Eastern Oyster was once one of the Outer Lands keystone species, both in its ecological role in shallow coastal waters and in how its abundance played a key role in nourishing both Native Americans and early colonists. In the Outer Lands virtually every sheltered bay and harbor area contained huge populations of oysters. Oysters are unusual bivalves in a number of ways. Most obviously, they don't bury themselves the way clams do: oysters are epifaunal creatures, and in normal circumstances they live in large crowds, or reefs, of shells that once carpeted the bottom of every river mouth along the coast. Each oyster shell is uniquely shaped, probably an adaptation to living in thick crowds of fellow oysters, where every inch of available space—no matter the shape—was valuable living room. Elongated, narrow oyster shells come from soft-

Atlantic Surf Clam
Spisula solidissima

Channeled Whelk
Busycotypus canaliculatus

Eastern Oyster
Crassostrea virginica

bottomed areas, but oysters that attach to firm surfaces tend toward a more rounded shape. Young oyster larvae are free-floating plankton, but they are highly attracted to chemicals that oysters give off and so tend to settle on hard surfaces near other oysters or on the shells of living or dead oysters. Oysters are unique in that they cannot move after they settle. Once the young oyster sets on a surface, it stays there for life.

Wellfleet Bay was once famous for its delicious—and superabundant—oysters, and the oyster industry was critical to the founding and early growth of the town of Wellfleet. Unfortunately, oysters were one of the first marine species to be overharvested, and even before the Revolution the oyster beds were gone, mostly from overharvesting but also because the shells of harvested oysters were ground and used by farmers to neutralize the acid soils of Cape Cod. This was a critical mistake for the oyster industry: we now know that young oysters must have a hard surface to set on to continue the renewal of the oyster reef, and young oysters have a strong instinct to settle on old oyster shells. With the increasing scarcity of shells, coupled with relentless harvesting, the Wellfleet Bay oyster reefs collapsed and disappeared, as did other smaller oyster reefs in the region. Efforts are now under way in Wellfleet Bay to reestablish oyster beds, and a small oyster fishery now exists, but it has been hampered by diseases and by the temperature of the shallows of Cape Cod Bay, which is warmer than it once was due to climate change.

Crabs and shrimp

Crabs are some of the most noticeable inhabitants of the subtidal zone, and they often range onto beaches and tidal flats, primarily at night. Three of the area crab species are active swimmers, although the Green Crab lacks the specialized swimming paddles seen on the legs of Blue Crabs and Lady Crabs. Blue Crabs, although numerous enough to be frequently caught by sport fishers, are not yet a major commercial fishery in southeastern New England as they are in the Chesapeake and Delaware Bays, but the recent warming of coastal waters has resulted in an increase in Blue Crabs. With the crash of American Lobster populations in southeastern New England over the past 15 years, the Blue Crab may eventually take over the ecological niche of bottom-dwelling predator-scavenger once held by lobsters.

Both Blue Crabs and Lady Crabs are active, aggressive predators that will catch and eat just about any kind of animal prey in the shallow subtidal zone. Both species should be handled with care to avoid a painful pinch, but the Lady Crab is particularly well known for its fast reflexes and strong claws.

The Green Crab is a nonnative species, one of the first major instances of a European animal that made the jump across the Atlantic, probably by riding on the mossy bottoms of sailing ships or in wet ballast stones dumped overboard in a Massachusetts port. Since 1817, the Green Crab has spread along the East Coast from Nova Scotia down to Cape May, New Jersey. Green Crabs are also voracious predators, particularly of small crabs, mussels, and Softshell Clams, but there may be a hidden positive side to Green Crabs: researchers from Brown University have discovered that salt marshes with Green Crabs are healthier because the Green Crabs prey on Purple Marsh Crabs, which eat marsh grasses.

The shells of these common subtidal crab species are often found along the wrack line of beaches, but that does not mean that many crabs have died or been killed by predators. Hard-shelled animals like crabs must molt their old shells to grow, and most crab species molt several times during the warm months. The molted carapace and claw shells that wash up on the shoreline are clean, with no organic matter inside, and are generally whole and in good condition. Crabs killed by gulls or herons are smashed and thoroughly dismembered, often on a rock or other hard surface, and you usually don't find the remains along the wrack line.

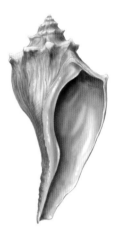

Knobbed Whelk
Busycon carica

In spite of its intimidating looks, the Common Spider Crab is a sluggish and inoffensive member of the deeper subtidal community, where it feeds on starfish and scavenges the remains of other bottom invertebrates. Spider Crabs prefer the deeper waters of the bays and harbors but are also common in shallow Eelgrass communities.

American Lobsters, formerly abundant in the subtidal zones of the Cape and Islands, are now mostly creatures of deeper bottom waters (see the next chapter, "Deeper Coastal Waters").

The Atlantic Horseshoe Crab is an unmistakable member of the tidal and subtidal communities, with its distinct shape, hard-leather shell, and menacing-looking (but harmless) long tail. Horseshoe Crabs are not true crabs at all but members of the ancient order Xiphosura, which also contains spiders and mites. They favor sandy or muddy bottoms and normally live at the deeper end of the subtidal zone, where they plow through the bottom, feeding on small invertebrates. In May and June, Horseshoe Crabs travel into the low tidal zone to lay eggs, often during a spring (unusually high) tide. These eggs are very attractive to shorebirds, and a good indication that the crabs are breeding is the sight of flocks of birds avidly picking at the eggs in the surf line. Commercial fishers

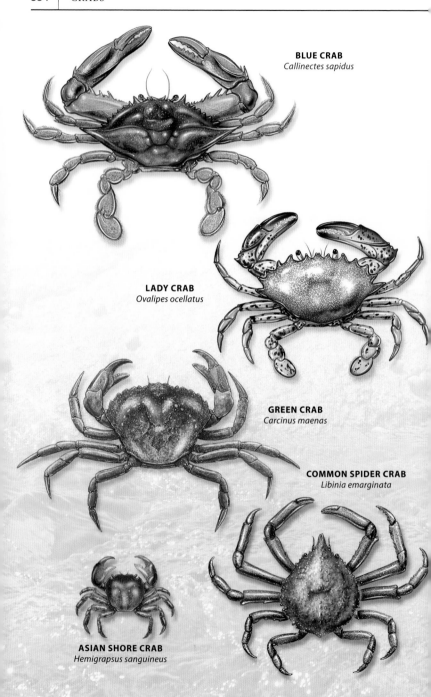

BLUE CRAB
Callinectes sapidus

LADY CRAB
Ovalipes ocellatus

GREEN CRAB
Carcinus maenas

COMMON SPIDER CRAB
Libinia emarginata

ASIAN SHORE CRAB
Hemigrapsus sanguineus

harvest Horseshoe Crabs for bait, and the Atlantic Coast population has fallen sharply over the past 40 years owing to overharvesting and the destruction of beach habitat.

Atlantic Horseshoe Crab blood is the sole source of an important medical compound—limulus amebocyte lysate (LAL)—used to test for the presence of harmful bacterial toxins in human blood. Drug manufacturers use LAL to test the safety of pharmaceutical and medical devices that contain blood products. Horseshoe Crabs are not usually killed in harvesting LAL. The crabs are caught, blood is drawn, and they are returned to the wild. About 60 percent of the crabs survive the experience.

Environmental groups monitor the population of Atlantic Horseshoe Crabs in various areas of southeastern New England, and you may see a round white identification tag attached to a shell. Please follow the instructions on the tag and report the tag number to the US Fish and Wildlife Service (http://www.fws.gov/crabtag/). The information is extremely useful in protecting populations of this valuable and threatened member of the shoreline community.

Atlantic Horseshoe Crab
Limulus polyphemus

Two other types of small crustaceans are commonly found in the subtidal zone: hermit crabs and grass shrimp (also called prawns). Colored a pale, translucent gray, grass shrimp are an important link in the estuary food chain. They feed on detritus—the remains of salt marsh grasses and other plants washed into bays and coastal waters. The nibbling of millions

Eggs (inset, lower left); baby crabs in early summer; and a tagged crab (inset, upper right). If you spot a tagged crab, please follow the instructions and report it; the research will help protect this species from overharvesting.

Photos courtesy of Frank Gallo.

Long-Clawed Hermit Crab
Pagurus longicarpus

Flat-Clawed Hermit Crab
Pagurus pollicaris

of grass shrimp breaks down plants into particles that become food for small zooplankton and bacteria, which complete the process of turning old marsh grasses into animal biomass. The shrimp are an important food source for larger predators such as crabs, the young of many fish species, and birds.

Hermit crabs are important scavengers on the subtidal zone, feeding on detritus and animal remains. Hermit crabs do not grow their own shells; instead, they adopt the discarded shells of snails for protection and shelter. The small Long-Clawed Hermit Crab is common in both rocky and sandy or muddy bottoms in the subtidal zone and usually inhabits old snail shells. The larger Flat-Clawed Hermit Crab prefers deeper waters with rocky or shell bottoms and usually picks larger homes such as old Moon Snail shells or small whelk shells.

Sea stars and sea cucumbers

Sea stars (starfish) and sea cucumbers are echinoderms, a phylum of animals whose bodies are arranged into five segments around a central axis. Common Sea Stars live in the subtidal zone and venture into the lower tidal zone in search of their favorite prey, clams and oysters. Common Sea Stars are a problem for oyster farmers because of their voracious appetite for oysters; they can infest and even wipe out entire oyster beds. Sea Cucumbers usually bury themselves in the bottom mud or sand to avoid predators, leaving only a ring of branching tentacles visible on the bottom surface.

Segmented worms and other bottom infauna

Along with the familiar clams, soft sand and mud bottoms contain a complex community of segmented worms known as the bristle worms or polychaetes. Clam worms and bloodworms are two of the most common types of larger worms and are familiar as bait for sport fishers. Both species move freely through the bottom sediments in search of small animal prey, including clams and other worms. Other marine worms are sessile, building permanent or semipermanent tubes in the mud from which they project their feeding appendages. Ice cream cone worms, bamboo worms, amphitrite worms, and feather duster worms are filter feeders, using their tentacles to grab plankton or small bits of organic material from the flowing water. Bottom worms are very sensitive to any vibration or unusual water movement near them and will quickly disappear into their tubes under the surface. The best tactic to observe bottom life is to find a likely spot in shallow, clear water, approach the area with care, and wait patiently until the bottom dwellers cautiously return to their normal feeding behavior.

Daggerblade Grass Shrimp
Palaemonetes pugio

Photo: Brian Gratwicke.

Fish

Small fish, including Atlantic Silverside, Striped Killifish, and Sand Lance, and the young of such larger fish as Bluefish and Striped Bass frequent the subtidal zone, particularly where vegetation or glacial boulders can shelter them from larger predatory fish. The subtidal zone and Eelgrass meadows are the great nursery areas of coastal waters, providing a wealth of food and protection for virtually all the major fish species found in the Outer Lands waters. Alewife, American Butterfish, American Shad, Atlantic Menhaden (Bunker), Black Sea Bass, Blueback Herring, Fluke, Scup (Northern Porgy), Smooth and Spiny Dogfish sharks, Tautog (Blackfish), and Winter Flounder are just some of the species that depend on the subtidal zone for a portion of their lives.

Birds

Many birds feed in the subtidal zone, even some you might not think of as seabirds. Long-legged waders including Great Blue Herons, Great Egrets, Snowy Egrets, Greater and Lesser Yellowlegs, and Glossy Ibis all stalk the shallows, picking off crabs, small fish, and other animals. Feeding Brant geese are most often seen floating over the subtidal zone, occasionally dipping down to feed on Eelgrass or Sea Lettuce.

Loons and grebes are diving birds that feed mainly on small fish in the subtidal zone but also take small clams and other bottom invertebrates. Common and Red-Throated Loons and Horned Grebes are frequent spring and fall migrants in New England coastal waters. The tiny Pied-Billed Grebe is a common migrant along the shore and major rivers in fall. It is less common in spring migration and is rare in the region the rest of the year.

One of the most abundant birds of the subtidal zone is the Double-Crested Cormorant, the angular black birds often seen sunning their outspread wings on docks, pilings, breakwaters, channel markers, and any other suitable perch along the shoreline. Cormorants are powerful underwater swimmers that feed primarily on small fish but will also take crabs when they can find them. In winter the larger and less common Great Cormorant may be seen, particularly on Cape Cod shores and harbor breakwaters.

Lesser Yellowlegs
Tringa flavipes

Common Sea Star
Asterias rubens

Illustrations not to scale; lengths cited are typical ranges

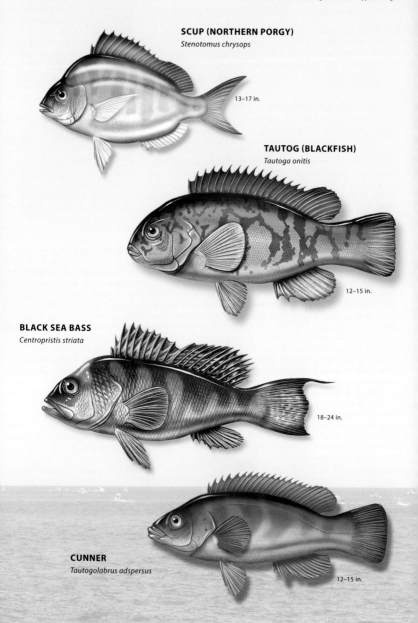

SCUP (NORTHERN PORGY)
Stenotomus chrysops

13–17 in.

TAUTOG (BLACKFISH)
Tautoga onitis

12–15 in.

BLACK SEA BASS
Centropristis striata

18–24 in.

CUNNER
Tautogolabrus adspersus

12–15 in.

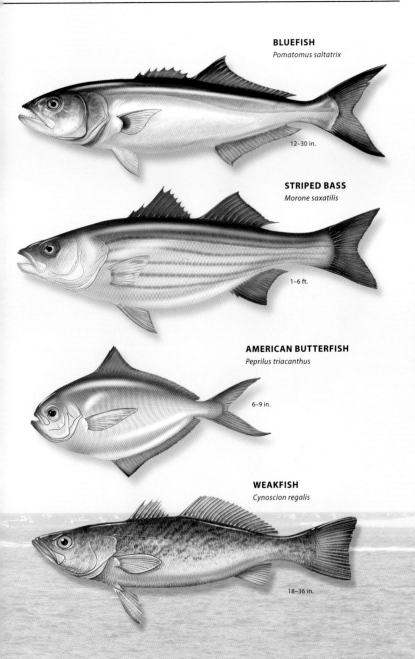

BLUEFISH
Pomatomus saltatrix

12–30 in.

STRIPED BASS
Morone saxatilis

1–6 ft.

AMERICAN BUTTERFISH
Peprilus triacanthus

6–9 in.

WEAKFISH
Cynoscion regalis

18–36 in.

Illustrations not to scale; lengths cited are typical ranges

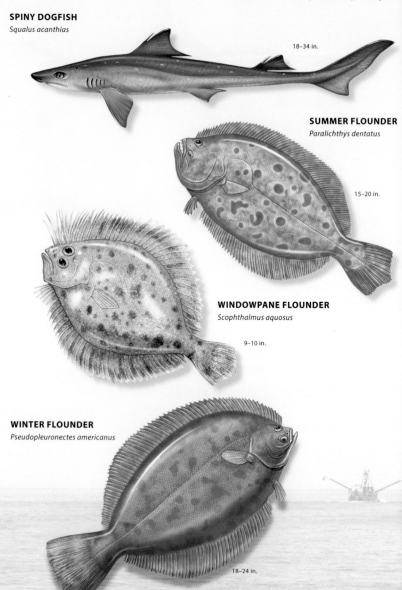

SPINY DOGFISH
Squalus acanthias

18–34 in.

SUMMER FLOUNDER
Paralichthys dentatus

15–20 in.

WINDOWPANE FLOUNDER
Scophthalmus aquosus

9–10 in.

WINTER FLOUNDER
Pseudopleuronectes americanus

18–24 in.

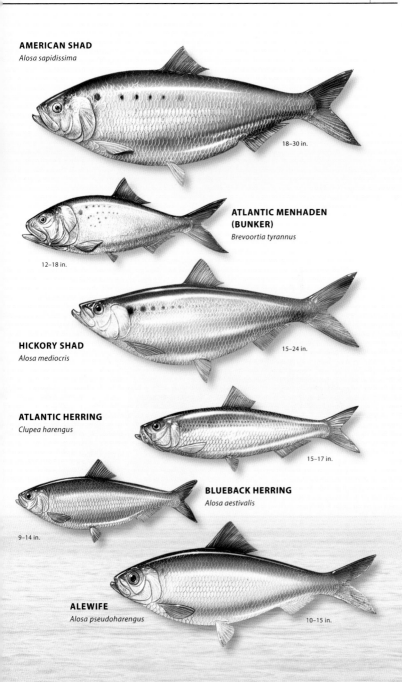

AMERICAN SHAD
Alosa sapidissima

18–30 in.

ATLANTIC MENHADEN (BUNKER)
Brevoortia tyrannus

12–18 in.

HICKORY SHAD
Alosa mediocris

15–24 in.

ATLANTIC HERRING
Clupea harengus

15–17 in.

BLUEBACK HERRING
Alosa aestivalis

9–14 in.

ALEWIFE
Alosa pseudoharengus

10–15 in.

Illustrations not to scale; lengths cited are typical ranges

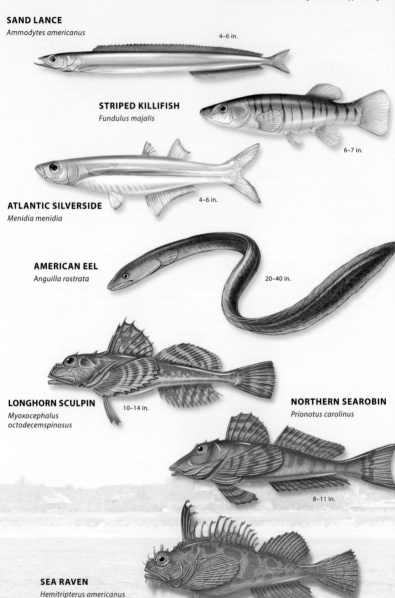

SAND LANCE
Ammodytes americanus

4–6 in.

STRIPED KILLIFISH
Fundulus majalis

6–7 in.

ATLANTIC SILVERSIDE
Menidia menidia

4–6 in.

AMERICAN EEL
Anguilla rostrata

20–40 in.

LONGHORN SCULPIN
*Myoxocephalus
octodecemspinosus*

10–14 in.

NORTHERN SEAROBIN
Prionotus carolinus

8–11 in.

SEA RAVEN
Hemitripterus americanus

9–22 in.

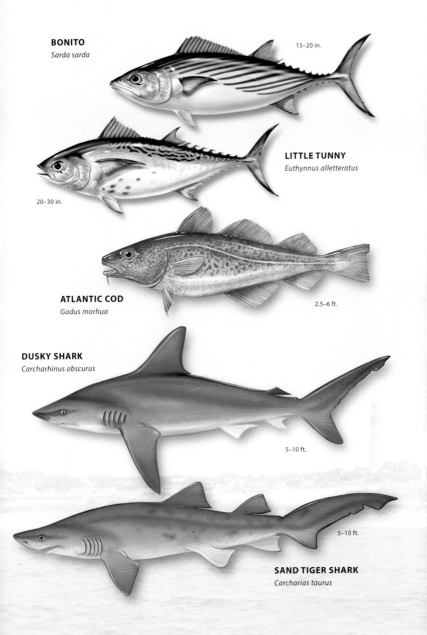

BONITO
Sarda sarda

15–20 in.

LITTLE TUNNY
Euthynnus alletteratus

20–30 in.

ATLANTIC COD
Gadus morhua

2.5–6 ft.

DUSKY SHARK
Carcharhinus obscurus

5–10 ft.

SAND TIGER SHARK
Carcharias taurus

5–10 ft.

Diving and dabbling ducks

In fall and winter the bird life of the Outer Lands coasts is largely defined by large rafts of duck species that all dive partially or fully underwater to feed. Dabbling ducks such as the Mallard and the closely related American Black Duck feed by tilting themselves downward from the surface, rarely fully submerging their buoyant bodies. Blue-Winged and Green-Winged Teals and the American Wigeon are also common dabbling ducks along the coast in spring and fall and, to a lesser extent, in winter. As dabblers, these ducks are limited to feeding on the immediate shoreline and to shallow, sheltered subtidal areas like harbors and bays, but when they are not actively feeding, the dabblers often drift well away from the shore for safety.

Pied-Billed Grebe
Podilymbus podiceps

The true diving ducks, which fully submerge and swim well underwater, are the most numerous ducks along the Cape and Islands coast in fall, winter, and early spring. Greater and Lesser Scaup, White-Winged and Surf Scoters, Common Goldeneyes, and Buffleheads are the most common diving ducks. All these species eat small bottom invertebrates and aquatic plants such as Sea Lettuce and Eelgrass.

Mergansers are diving ducks whose long, thin bills are edged with toothlike serrations that help them seize their specialty: small fish and slippery aquatic invertebrates. Red-Breasted and Hooded Mergansers are common in the cold months but leave the Outer Lands area to breed in freshwater lakes and ponds in more northern areas of the United States and Canada.

Gulls, terns, and skimmers

Double-Crested Cormorant
Phalacrocorax auritus

The larger gull species of the region are numerous and aggressive predators along the immediate shoreline. Gulls usually do not dive below the water surface for food and thus don't feed directly in the subtidal zone, but they nevertheless benefit from the rich pickings of subtidal creatures like crabs, marine worms, and clams that are at or just below the low tide line. Gulls are discussed more fully in the chapter on beaches, but the most common gulls of the Outer Lands are the Herring Gull and Ring-Billed Gull.

Terns are smaller relatives of gulls and are typically white with a black cap, long, black-tipped wings, and a long, forked tail. As its name suggests, the Common Tern is the tern most frequently seen over subtidal waters, where it dives for small fish captured at or near the water surface. A number of areas along the Outer Lands coasts are protected during nesting season to help rebuild the Least Tern population, which has been severely reduced over the past century by the loss or disturbance of their breeding habitat: sandy beaches. Least Terns also feed by making shallow dives for small fish. Looking for groups of feeding terns is a great way to spot schools of small fish being chased by larger predators like Bluefish. Terns will hover over the feeding frenzy, making quick dives that rarely take them below the surface. The tern's quick snatch-and-grab diving method may be a matter of safety as much as efficient feeding. Hungry Bluefish do not discriminate between fish and birds, and a surprising number of terns have lost a lower leg or foot to a voracious Bluefish.

The Black Skimmer is a close relative of terns with a spectacular method of feeding: skimmers fly just above the surface of subtidal waters, skimming for small fish by dipping their oversized lower bills into the water and snapping them shut the instant that they hit a fish. A few decades ago Black Skimmers were an unusual August and September treat for birders along Cape and Islands coasts, as skimmers are a more southern species. As the climate has warmed, Black Skimmers

Red-Breasted Merganser
Mergus serrator

GREAT BLUE HERON *Ardea herodias*

SNOWY EGRET *Egretta thula*

kreefax

GREATER YELLOWLEGS *Tringa melanoleuca*

LESSER YELLOWLEGS *Tringa flavipes*

PIED-BILLED GREBE *Podilymbus podiceps*

MUTE SWAN *Cygnus olor*

ALLARD *Anas platyrhynchos*

Jim Shane

MERICAN WIGEON *Anas americana*

AMERICAN BLACK DUCK *Anas rubripes*

Erni

BUFFLEHEAD *Bucephala albeola*

Steve Byland

LUE-WINGED TEAL *Anas discors*

Erni

GREEN-WINGED TEAL *Anas crecca*

Marco Barone

BRANT Branta bernicla

feathercollector

COMMON GOLDENEYE Bucephala clangula

Karen Popovich

GREATER SCAUP Aythya marila

Steve Byland

LESSER SCAUP Aythya affinis

M. Carter

SURF SCOTER Melanitta perspicillata

Paul Reeves Photography

WHITE-WINGED SCOTER Melanitta fusca

have become regular late summer visitors, and a few pairs of skimmers have nested in southern New England in the recent past. Skimmers are fairly common on the southern shore of Long Island, where they breed in tern colonies.

Seals and White Sharks

As recently as the 1980s Gray Seals and Harbor Seals were regular but not very noticeable residents of Cape and Islands shorelines, usually seen in isolation or in small herds on isolated beaches such as Monomoy Island south of Chatham or Long Point off Provincetown Harbor. Since the 1980s Cape Cod, Nantucket, and especially Monomoy Island and Muskeget Island off Nantucket have experienced an explosive growth of Gray Seal colonies.

The sparse seal populations of the late twentieth century were no natural accident. From 1888 to 1962, Massachusetts paid a bounty of five dollars per head for killing seals, and during the program somewhere between 73,000 and 137,000 seals were killed by hunters in the Cape and Islands region. The Marine Mammal Protection Act of 1972 made killing seals a federal crime, and both legal and illegal killing of Gray and Harbor Seals ceased in the area. The huge growth in seal populations in the past 30 years should thus be understood as species

Hooded Merganser
Lophodytes cucullatus

resident in the area for tens of thousands of years reassuming their former populations and ecological role in the region.

Although all seal species in the Outer Lands are much more common than they were three decades ago, the huge growth of the Gray Seal population has received the most attention. Solid scientific estimates of the Gray Seal population have not yet been made because it's expensive and time-consuming to count such large numbers of marine mammals, but rough estimates are that some 30,000–50,000 Gray Seals inhabit the region. Gray Seals are present year-round, and their diet includes Striped Bass, Atlantic Cod, Bluefish, Atlantic Menhaden, and large numbers of a much tinier but important forage fish called Sand Lance. Sand Lance are eaten by a wide variety of wild animals, including many seabirds as well as Humpback Whales.

The growth in the seal population has been controversial on the Outer Cape, pitting the large tourist, natural history, and whale-watching communities against local commercial fishers, who see the seals as threats to the dwindling stocks of regional commercial and game fish. Environmentalists point

Gray Seals
Halichoerus grypus

out that although the seals certainly have an effect on local fish stocks, their impact is tiny compared to the decades of overfishing that brought about the sharply diminished New England commercial fishing industry.

The seals seen in Cape and Islands waters are part of a larger Gray Seal population that includes the entire Gulf of Maine and a huge breeding colony on Sable Island, south of Nova Scotia, home to nearly 400,000 Harbor and Gray Seals. Thus the discussion of "what to do about Gray Seals" is pointless: even if the current Gray Seal population of the Cape and Islands were to disappear overnight, in just a few years the much larger Sable Island herd would repopulate the excellent habitat offered by Monomoy Island, Muskeget Island, and the beaches of the Outer Cape.

The Great White Shark

The explosive growth of the Gray Seal population in the Cape and Islands region has drawn the attention of Great White Sharks, one of the ocean's most powerful predators. Cape Cod has become one of the few known aggregation sites for Great Whites in the Western Atlantic. The presence of Great White Sharks has built rapidly in the past decade. In 2012 there were 21 reported sightings of Great Whites. In 2017 there were 147 Great White Shark sightings from spotter planes over the Outer Cape and Monomoy Island, and by 2016 more than 100 Great White Sharks had been tagged as part of the ongoing Massachusetts Division of Marine Fisheries Shark Research Program.

Great White Sharks are visual ambush hunters. They typically cruise low in the water column, rushing upward when they spot a seal silhouette near the surface in order to deliver a massive killing bite, and then backing off while their victim bleeds to death before moving in to feed.

The threat posed by Great White Sharks has changed the behavior of Cape and Islands Gray Seals. In the warmer months of the year, when sharks are present, the seals spend most days on the beach, only moving off the beach to feed in the dawn and dusk

Harbor Seal
Smaller, with a puppylike, rounded head profile

Gray Seal
Larger, with a thick, horselike snout

twilight hours and at night, when the sharks have a tougher time spotting the seals. This recent tendency for the seals to spend the daylight hours out of the water has made interactions between human beachgoers and seals more frequent, particularly on the ocean beaches of the Outer Cape from Chatham north to Race Point.

Ice seals

Outer Lands beaches are now regularly visited by three species of seals that normally live and winter north of the Saint Lawrence River off Newfoundland and Labrador. Sightings of these so-called ice seals are unusual, and mostly happen only in isolated beaches in winter, such as those on the Outer Cape, Monomoy Island, and the stretch between Race Point and Long Point in Provincetown. But each passing year brings more scattered reports of sightings and strandings of these rare seals.

The Harp Seal and the Hooded Seal are the more regularly spotted ice seals. The more unusual Ringed Seal is known from scattered observations and strandings. It is not clear why these northern seal species are more often wandering south of their normal ranges, but most of the individuals spotted or stranded are juveniles, not adults.

West Indian Manatee

Another oddity spotted a number of times in recent years is the West Indian Manatee, which normally is not seen in any numbers north of the Georgia coast. However, over the past decades isolated individual manatees have been seen in various locations along the East Coast as far north as Cape Cod. In the most recent incident, a young female manatee was sighted in the waters off Chatham in August 2016 and captured in September 2016 near Falmouth because the animal was believed to be cold-stressed and would certainly not have survived the New England winter. The manatee was examined at Connecticut's Mystic Aquarium and later released off the Florida coast.

Manatees are known to wander north in small numbers with the warming waters in summer and early fall. These rare, isolated incidents are probably due to the energy and persistence of individual manatees and do not seem to be related to global warming or other climate changes. Manatees feed primarily on sea grass in shallow, subtidal waters, and unfortunately the relatively small remaining Eelgrass meadows of the Outer Lands offer little suitable food.

Harbor Seals
Phoca vitulina

HARBOR SEAL
Phoca vitulina

Harbor Seal
pup

GRAY SEAL
Halichoerus grypus

The pelage colors
and patterns of
young Gray Seals
vary from almost
pure white to yel-
low or gray.

GRAY SEAL

HARBOR SEAL

Mark Bridge

Wim Claes

HARP SEAL
Pagophilus groenlandicus

HOODED SEAL
Cystophora cristata

Female

Male

RINGED SEAL
Pusa hispida

Manatees are rarely seen north of South Carolina, but every few years a single (usually young) Manatee wanders north into southern New England waters.

WEST INDIAN MANATEE
Trichechus manatus

Eelgrass communities

Eelgrass beds develop in shallow-water areas with soft bottom sediments. Eelgrass is widespread all along the coasts of the North Atlantic Ocean and is the dominant sea grass species north of Cape Hatteras on the East Coast. Eelgrass is a flowering grass (family Zosteraceae) that has adapted to life in salt water. It spreads primarily through underground stems called rhizomes within the bottom sediments. The thick tangle of Eelgrass rhizomes stabilizes soft sediments and keeps the plants from washing away.

Plants and animals of Eelgrass meadows

Healthy Eelgrass beds offer both food and shelter to the young of many fish species, as well as adult fish. Atlantic Silverside, Spot, Tautog (Blackfish), and Summer Flounder all find shelter in Eelgrass meadows. The many small fish also attract predators: Bluefish and Striped Bass cruise the meadows in search of prey, and many bird species find food in the Eelgrass. Dabbling ducks, geese, and swans eat the grass directly. Diving ducks such as the Red-Breasted Merganser and Bufflehead pick off young fish, shrimp, snails, and worms living within the Eelgrass. In shallower Eelgrass areas, long-legged waders such as the Great Blue Heron and the Great Egret stab for fish and shrimp in the meadows. Historically Eelgrass was the primary food of Brant, a common small goose that winters in Long Island Sound and along ocean coasts.

Eelgrass is not the only marine flowering plant species in the Outer Lands. Widgeon Grass also grows in Eelgrass beds and is an important food source for many diving and dabbling duck species (a wigeon—slightly different spelling—is a kind of dabbling duck). The leaves of Widgeon Grass are much more slender than those of Eelgrass, and they are eagerly sought by American Wigeons, American Black Ducks, scaup, teals, and other coastal ducks. Sea Lettuce is usually abundant in Eelgrass beds as well.

Eelgrass beds are a crucial habitat for the Bay Scallop. The young scallops attach themselves to Eelgrass stems well above the bottom, and this protects them from predatory crabs. Eelgrass beds slow the currents of water within and over them. This calming of currents, and the complex structure created by all the grass leaves, provides valuable shelter for many small invertebrates. Daggerblade Grass Shrimp are common in Eelgrass beds, where they form an important food resource for the many species of young fish that shelter in Eelgrass meadows. The beds also shelter bivalves like the Northern Quahog, which are present in large numbers in healthy Eelgrass communities.

Brant
Branta bernicla

NORTHERN PIPEFISH
Syngnathus fuscus

ATLANTIC SILVERSIDE
Menidia menidia

TAUTOG (BLACKFISH)
Tautoga onitis

BAY SCALLOP
Argopecten irradians

BLUE CRAB
Callinectes sapidus

NORTHERN QUAHOG
Mercenaria mercenaria

COMMON SEA STAR
Asterias rubens

LINED SEAHORSE
Hippocampus erectus

BAY SCALLOP
Argopecten irradians

ATLANTIC HORSESHOE CRAB
Limulus polyphemus

The head and open nostrils of a Humpback Whale surfacing to breathe after a deep dive on Stellwagen Bank off the northern tip of Cape Cod.

DEEPER COASTAL WATERS

A classic tail roll of a Humpback Whale angling downward for a deeper and longer feeding dive.

The deeper coastal waters around the Outer Lands are home to an amazing array of animals. This region is justly famous for its world-class whale watching and sport fishing, but the whales, seabirds, seals, sharks, and fish are here because of the abundance of food. The complex interactions of the sun, clashing ocean currents, undersea landscape, and mobile animals that range from the near-tropical waters of the Gulf Stream 200 miles to the south all the way to the icy waters of the central Gulf of Maine are the sources of this natural wealth. At Stellwagen Bank (see map, p. 399) and Georges Bank (see map, pp. 4–5) in particular, the physical and bio-logical elements combine to produce areas of extraordinary beauty and bounty that in turn enrich the whole region.

Bottom environments

The deep bottom environments offshore are diverse, ranging from pure sand to a mix of silt and sand, shells, and small stones to hard rock piles scattered with glacial boulders erod-ed from the soft clays of cliffs and moraines created by the Wisconsinan ice sheet. Each bottom environment supports a different mix of animals. The softer bottoms, for example, support a rich infauna of worms and other marine inverte-brates that are largely absent on hard bottoms. The depths lack enough light for large, attached algae to grow, so there are no fixed plants. Wave action is also not a factor in the depths, although strong currents can sweep through, particularly near tight passages between landforms, as in the notorious Pollock Rip between the south end of Monomoy Island and Great Point in Nantucket.

ANIMALS OF OFFSHORE WATERS AND STELLWAGEN BANK

HUMPBACK WHALE
Megaptera novaeangliae

GREAT WHITE SHARK
Carcharodon carcharias

BLUE SHARK
Prionace glauca

BASKING SHARK
Cetorhinus maximus

OCEAN SUNFISH
Mola mola

FORAGE FISH SPECIES
Smaller animals; images not to scale

BLUEBACK HERRING
Ammodytes americanus

ATLANTIC SILVERSIDE
Menidia menidia

ATLANTIC MENHADEN
Brevoortia tyrannus

SAND LANCE
Ammodytes americanus

ATLANTIC HERRING
Clupea harengus

LONGFIN SQUID
Doryteuthis pealeii

Deep bottom communities

Many of the invertebrates that live in sand or muddy areas of the subtidal community also inhabit the depths. In particular, ice cream cone worms, bamboo worms, bloodworms, and clam worms live in deeper soft bottom areas, as do many clam species familiar from shallower waters. Several species of crabs are common in the deep bottom community, including the Common Spider Crab and the similar-looking Rock and Jonah Crabs. The Rock Crab is widely distributed across all bottom types, but the Jonah Crab is more common on rocky bottoms.

Among the elasmobranchs (boneless sharks, skates, and rays), the common Smooth and Spiny Dogfish are joined by Little Skates, Barndoor Skates, and Winter Skates, all of which hunt clams, crabs, and other bottom invertebrates and small fish over soft sediment bottoms.

The shallows of the continental shelf off the Outer Lands, as well as the various sounds, bays, and inlets, support a surprisingly varied community of bottom fish species. Winter Flounder and Windowpanes are common flatfish in the depths. The deep bottom community also supports a range of oddly shaped searobin and sculpin relatives, such as the Northern Searobin, Sea Raven, and Longhorn Sculpin (see the previous chapter, "Shallow Coastal Waters," for fish illustrations).

The American Eel is an important and common member of the bottom community, particularly near river mouths and in brackish water. American Eels are nocturnal. They hide in crevices or burrow into sand during the day and hunt at night. The American Eel is the region's best-known catadromous fish species (living in freshwater but spawning in salt water).

Jonah Crab
Cancer borealis

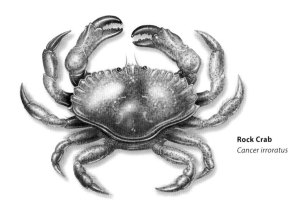

Rock Crab
Cancer irroratus

Though still numerous, American Eels were once a much more common member of both the freshwater and saltwater communities around the Outer Lands, but the lack of large rivers and the construction of dams and other obstructions to migration have resulted in reduced eel numbers.

Rocky bottom communities

Over rocky bottoms the water is generally clearer and freer of the silt that can smother many types of marine filter feeders. Jonah Crabs favor rocky bottoms, but the closely related Rock Crab is also common. Fish that specialize in rocky bottom areas include Cunner, Tautog (Blackfish), and Scup (Northern Porgy). The American Lobster is the most famous and economically valuable resident of the Gulf of Maine's rocky bottom communities.

Barndoor Skate
Dipturus laevis

American Lobster

Thanks to its delicious meat, the American Lobster is known far and wide as a symbol of coastal New England. A decapod (10-legged) crustacean in the same family as shrimps and crabs, the lobster shares the same basic body plan, but with massively enlarged claws and a stretched abdomen with a powerful muscular tail. The lobster's large claws are asymmetric: the larger crusher claw is used for cracking the shells of sea urchins and mollusks, and the finer cutter claw is used for extracting meat and more delicate maneuvering of prey. Before they became relentlessly hunted, lobsters could live three decades or more. Today most inshore lobsters rarely reach six years of age before being caught.

Lobsters can live almost anywhere in the rocky areas offshore. Because they are nocturnal and dislike strong light, lobsters are rarely seen in shallow water. During the day they hide

Winter Skate
Leucoraja ocellatus

American Lobster
Homarus americanus

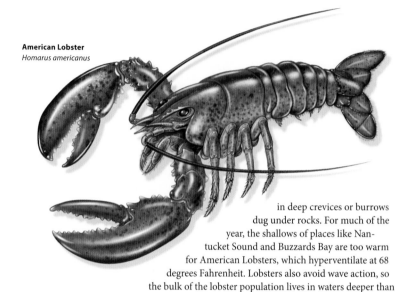

in deep crevices or burrows dug under rocks. For much of the year, the shallows of places like Nantucket Sound and Buzzards Bay are too warm for American Lobsters, which hyperventilate at 68 degrees Fahrenheit. Lobsters also avoid wave action, so the bulk of the lobster population lives in waters deeper than 15 feet.

Lobsters roam the rocky bottom areas in the dark, foraging for and feeding on clams, scallops, sea urchins, and other living or dead animal material they can find. Lobsters are powerfully attracted to the scent of fish flesh, which is why lobster traps baited with chunks of fish work so well, but lobsters prefer live, healthy food over decaying animals. Lobsters are very mobile and make long migrations across the undersea landscape. Migrating American Lobsters can travel along the ocean bottom more than a mile every day, and much of the Outer Lands lobster population migrates each autumn to winter in deeper waters.

Lobster populations in New England

The American Lobster is a cold-water crustacean. Although lobsters live along the East Coast from Labrador to Virginia, the major populations are in the northern Gulf of Maine and around the Gulf of Saint Lawrence. The southern waters of the Outer Lands and Long Island Sound have always been the southern limit of inshore lobster populations; farther south, the lobster is strictly a deep-water animal never found near shore. Lobster populations off Long Island, Rhode Island, Block Island, and in Long Island Sound have undergone a spectacular crash since 1999, even as Massachusetts and Maine catches have broken records for abundance.

Most experts think that warmer waters in southeastern New

England will cause a permanent commercial extinction of the American Lobster south of Cape Cod, and that disease and pollution effects are just the most visible manifestations of a simple fact: most Outer Lands waters are becoming too warm to support a significant population of lobsters. Even the cooler waters of the southern Gulf of Maine are warming, and lobster fishers are wary of what the next decade will bring in the southern Gulf of Maine and Cape Cod Bay.

Open water

As every fisher knows, open-water wildlife have a patchy distribution across the marine environment. One moment the sea seems to be boiling with predators and prey fish, and moments later all the fish seem to have disappeared without a trace. In this zone the most important food sources for sea-birds, whales, seals, and the larger game fish are plankton (for North Atlantic Right Whales and Basking Sharks) and small-to-medium-sized schooling fish such as Atlantic Menhaden, Sand Lance, Atlantic Herring, and American Butterfish. The Longfin Squid is an important food species for both predatory fish and seals.

NOAA

Planktonic larva stage of the American Lobster (*Homarus americanus*).

In warm months the coastal waters of the Outer Lands are one of the most important nursery areas for coastal commercial and sport fishing species. In winter the regional waters are a major East Coast refuge and larder for a large population of wintering waterfowl and coastal birds.

In the deeper coastal waters there are two major biological components: plankton and nekton. Plankton are tiny plants and animals that swim weakly or passively drift with the tides and currents. Nekton are larger, stronger swimming animals such as open-water fish species, but also include other power-ful swimmers like squid and marine mammals. Plankton are divided into two major groups: phytoplankton and zoo-plankton. Phytoplankton are single-celled plants that create biomass through photosynthesis, and zooplankton are tiny animals, including the larvae of many fish and invertebrates. The life cycle of most offshore animals includes both a plank-tonic and a nektonic phase: they hatch as tiny planktonic larvae and later mature as freely swimming nekton.

Plankton

Phytoplankton and zooplankton collectively make up the critical food resource for the whole open-water food chain. The population densities of plankton are almost beyond imagining. At their densest, in late winter, phytoplankton may be as much as 400 green cells per quart of water. Zooplankton density is about 200 copepods, or immature stages, of fish,

DIATOMS
Asterionellopsis glacialis

DIATOMS
Chaetoceros sp.

DIATOMS
Thalassionema frauenfeldii

DIATOMS
Odontella sinensis

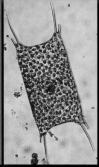

Not to scale and highly magnified over life size

All plankton images: NOAA Photo Library.

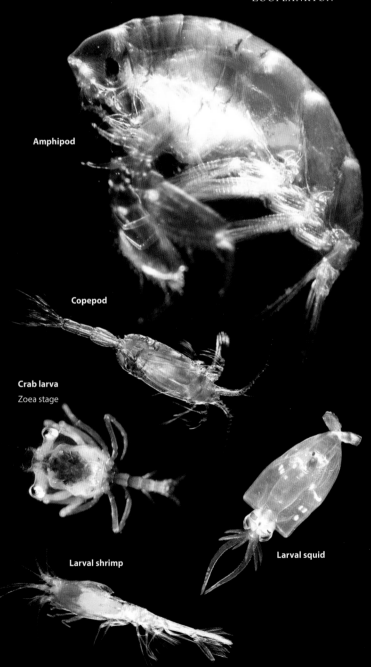

Amphipod

Copepod

Crab larva
Zoea stage

Larval shrimp

Larval squid

Not to scale and highly magnified over life size

Atlantic Sea Nettle
Chrysaora quinquecirrha

shrimp, crabs, and mollusks per quart of water in late summer.

Even though phytoplankton are critical to life in coastal waters, in the past few decades the Outer Lands coastal waters have often suffered from too much of a good thing. Algae and other phytoplankton populations are ordinarily held in check by the natural scarcity of nitrogen, a vital nutrient for plant growth. Today, however, excessive nitrogen enters the water supply, chiefly from the effluent of water treatment plants and untreated runoff entering street drainage that both flow into coastal waters. The excess nitrogen encourages abnormal amounts of algal growth, which in turn depletes dissolved oxygen. Although algae, like all green plants, emit oxygen as part of photosynthesis, they also consume oxygen at night when they are not photosynthesizing. As the algae population explodes in the warm waters of summer, oxygen levels in smaller bays and inlets all around the Outer Lands can fall dramatically, and they occasionally drop to levels too low to sustain marine life, resulting in mass fish kills.

Phytoplankton

Phytoplankton are the most numerous organisms in coastal waters. More than 200 species of single-celled diatoms, dinoflagellates, green algae, and blue-green algae are the primary nutrient producers of the marine food chain. Most phytoplankton are diatoms, which have silica shells in complex geometric shapes that are often linked together to make clumps or chains of individuals. Diatoms are true plankton that drift with currents and tides. Dinoflagellates have whiplike cilia that allow them to move through the water for short distances and are most abundant in the early summer months.

Zooplankton

Zooplankton consist largely of minute animals of various kinds, such as copepods and tiny shrimp, that spend their whole lives drifting in surface waters with the currents. Other zooplankton, however, consist of the larvae of various invertebrates and fish that spend just a portion of their lives drifting as plankton. Most zooplankton are too tiny to be strong swimmers, but the zooplankton layer itself ascends and descends daily in the water column. In the night hours it rises to near the surface, and during the day it descends into the relative gloom of deeper waters, probably to lessen the threat of predators.

Most people don't normally think of Blue Crabs, starfish, Eastern Oysters, American Lobsters, or barnacles as free-floating residents of open coastal waters, but during their larval stages, these and many other bottom invertebrates are zooplankton. Many fish species also hatch as larvae before they reach a size where they can swim effectively. Normally sessile or slow-moving creatures like oysters or crabs spread their offspring far and wide as free-floating larvae that later settle in suitable territories for the benthic, or attached, phase of their lives. Most of the trillions of larvae hatched each year don't live more than a few days or weeks, instead becoming food for larger plankton and small fish. In this way, the larvae are a vital link in the marine coastal food chain.

Pelagic invertebrates

Larger pelagic invertebrates like sea jellies, comb jellies, and squid consume large amounts of zooplankton. Comb jellies (Ctenophores) are simple sea jelly–like animals that are sufficiently different from true sea jellies to be placed in their own phylum. Their common name comes from the rows of beating cilia arranged in long lines, or combs, across their surface, which allow these animals to swim and maneuver. Comb jellies feed on zooplankton and can consume up to 10 times their weight in prey each day. Masses of comb jellies can temporarily deplete the zooplankton of harbors and bays. Comb jellies are harmless to humans and can be safely handled because they do not have stinging cells. At night, comb jellies give off a faint green bioluminescence when disturbed by sudden water movement.

Evan Travels

Beroe Comb Jelly
Beroe sp.

The Sea Walnut is the most common comb jelly in coastal waters. The Sea Walnut is present year-round but is most common in late summer and early fall. A less common relative, Beroe Comb Jelly, is occasionally seen in offshore waters.

True sea jellies (Cnidarians) have a swimming form called a medusa with a pulsing bell and dangling tentacles equipped with stinging cells called nematocyts. All sea jellies go through a complex multistage life cycle, in which free-floating planktonic larvae settle on fixed bottom surfaces and grow into a polyp stage that releases small, free-floating medusae (tiny sea jellies) that then grow into the forms we see swimming in coastal waters. All sea jellies should be approached with caution, particularly until you have confidently identified the species.

Moon Jellies become common in late spring and become less widespread after mid-July. Their milky, translucent bells typically reach five to eight inches in diameter and often wash up on beaches. Moon Jellies have stinging cells on their short tentacles, but they are not numerous or strong, and most people have either a mild rash from contacting the tentacles or no reaction at all. Some people have strong allergic reac-

Longfin Squid
Doryteuthis pealeii

Jiang Zhongyan

**CANNONBALL
JELLY**
*Stomolophus
meleagris*

LION'S MANE JELLYFISH
Cyanea capillata

Seen from above,
as in shallows
or on a beach

Credits– Cannonball Jellies: ymgerman, Sky2015; Lion's Mane Jellyfish: Greg Amptman, Konstantin Novikov; PMOW: MSNN, sciencepics;
Atlantic Sea Nettle: Gino Santa Maria; Comb Jelly: John Wollwerth; Moon Jellies: Hans Hillewaert.

Portuguese Man o' Wars are sea jelly–like animals that appear sporadically in regional waters, usually in late summer. These jellies have a powerful sting and very long tentacles, so stay well away from them. The tentacles can sting long after the animal has died or has washed up on the beach.

PORTUGUESE MAN O' WAR
Physalia physalis

MOON JELLY
Aurelia aurita

ATLANTIC SEA NETTLE
Chrysaora quinquecirrha

COMMON NORTHERN COMB JELLY
Bolinopsis infundibulum
(A ctenophore, not a true jelly)

Not to scale

tions to any sea jelly venom, however, so it's best to avoid handling sea jellies.

The Lion's Mane Jellyfish is the large, red-violet, sometimes dinner-plate-sized sea jelly most often encountered in the shallow waters off beaches in late summer and early fall. The Lion's Mane is widely distributed in the North Atlantic and Pacific Oceans, and in Arctic waters it can grow to a diameter of seven feet or more, making it the world's largest sea jelly. In our area, however, these jellies rarely reach bell diameters beyond 12 inches. The Lion's Mane has very long tentacles with strong stinging cells, so do not approach one closely while swimming, and avoid stepping near any beached bells with bare feet.

The smaller Atlantic Sea Nettle is less common than other sea jellies but worth watching out for because its sting is so painful. Sea Nettles tolerate very low salinities and may be present in river mouths and harbors in late summer. The Atlantic Sea Nettle's bell is about the size of a Moon Jelly's bell, but the Sea Nettle has much longer, dark red stinging tentacles trailing the bell, so always give it a wide berth.

Squid
Although most bathers, sport fishers, and boaters rarely notice it, the Longfin Squid is quite common from late spring through early fall. Coastal waters and bays are important nursery areas for this widespread Atlantic Coast species, and

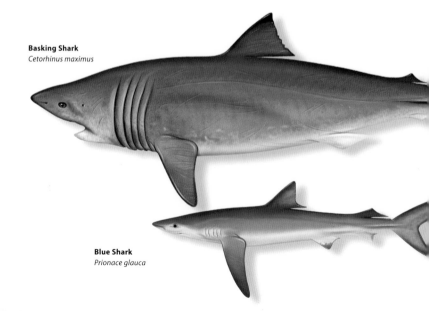

Basking Shark
Cetorhinus maximus

Blue Shark
Prionace glauca

NOAA

these strong open-water swimmers are a major food resource for Bluefish and Striped Bass, as well as Harbor and Gray Seals. Like most squid species, the Longfin Squid can rapidly alter its skin color and pattern by changing the size and shape of special pigment cells in its skin.

Longfin Squid
Doryteuthis pealeii

Most adult Longfin Squid are about 12–16 inches in length. Squid have a relatively short life cycle for an animal their size; adults live for only about a year. Squid breed primarily from May through September. Each female lays a large bundle of egg capsules, called a mop, in shallow areas, often attaching the mop to rocks or the fronds of brown algae for protection. Most eggs hatch in September, and small, one-to-two-inch squid are abundant during the fall months.

Fish
Through the year there are major shifts in species diversity and abundance in the coastal waters. More than 120 species of fish spend at least part of their lives in the region, and 50

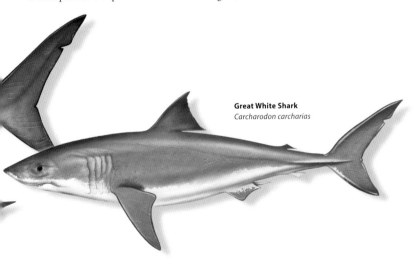

Great White Shark
Carcharodon carcharias

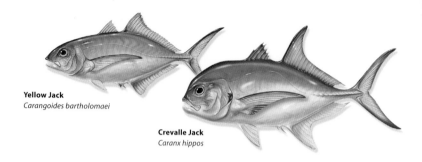

Yellow Jack
Carangoides bartholomaei

Crevalle Jack
Caranx hippos

species breed in the coastal waters of the Outer Lands. Overall fish abundance is highest in the late summer and early fall, but groundfish (fish that live on or near the bottom) are more numerous in the spring and decline through the summer, whereas open water fish increase through the summer and are most abundant in autumn (see illustrations, pp. 338–43).

In spring the most abundant groundfish species are Winter Flounder, Windowpane Flounder, Fourspot Flounder, Tautog (Blackfish), Little Skate, Red Hake, Scup (Northern Porgy), and Smooth and Spiny Dogfish. The most abundant spring open water fish species are schools of Atlantic Herring. Schools of Atlantic Menhaden enter the coastal waters in late spring, as do groups of the anadromous species American Shad, Blueback Herring, Alewives, and Atlantic Salmon, heading for rivers in their sadly reduced spawning runs. As spring progresses into summer the numbers of Bluefish, Striped Bass, and other midwater game fish steadily build. Weakfish move into the region from more southern waters in midspring, though in diminished numbers in recent years.

By late summer Scup (Northern Porgy) and Pelagic Butterfish are numerous in the region. Large schools of Atlantic Menhaden and their predators, Bluefish, roam throughout Outer Lands waters. In late summer warm-water and pelagic fish wander into coastal waters, including filefish, Crevalle Jacks, Yellow Jacks, Bonitos, Little Tunny, and Atlantic Spanish Mackerels. Schools of large Cownose Rays and Bullnose Rays may appear in late summer. Large populations of smaller fish species build through the summer. Atlantic Silverside, Sand Lance, and other small fish species are a crucial link in the estuary food chain and are the main food species for small gulls, terns, cormorants, skimmers, and many other water birds, as well as dolphins, whales, and seals.

In autumn most of the large schools of midwater and deep-water fish leave the region as the water cools. By late fall most

Bluefish and other game fish have moved out to deeper waters farther south along the East Coast. Some Striped Bass linger in harbors and river mouths until early winter.

Sharks

Three larger shark species are notable. Most often seen at the surface is the Basking Shark. These large sharks are filter feeders, and as their name suggests, they feed slowly at the surface on tiny copepods and other plankton. Basking sharks can look alarming because of their size and large triangular fins, but they are completely harmless unless harassed by boaters, and then the danger is from the powerful tail. Basking Sharks do not bite.

The Great White Shark has received massive media attention due to its reputation as one of the ocean's top predators, drawn to Cape Cod waters by the exploding numbers of Gray Seals in the Chatham, Nauset Beach, and Monomoy areas. The Great Whites typically arrive in the Cape area in mid-June and are present in numbers to early November. Massachusetts researchers currently have a five-year study under way to tag and track Great Whites, both to study the biology of the sharks and to develop a warning system for bathers along the beaches of the Outer Cape. In offshore waters Great Whites are rarely seen, as they tend to cruise well below the surface. Fishers handling a lot of bloody cleanings may attract their attention, but this is rare in Outer Lands waters.

Sport and commercial fishers handling or cleaning catches in offshore waters are sometimes visited by large, sleek Blue

The magnificent Atlantic Bluefin Tuna (*Thunnus thynnus*) is one of the most endangered species in the ocean, yet in a craven political decision in 2011, the NOAA fisheries division refused to add the Bluefin to the US Endangered Species Act. The widely respected IUCN Red List of Threatened Species lists the Atlantic Bluefin status as Endangered, largely due to overharvesting by commercial and private fishers.

ATLANTIC BLUEFIN TUNA
Thunnus thynnus

In recent years hundreds of cold-stunned sea turtles have washed up on the shores of Cape Cod. Over 90 percent of the turtles found on Cape beaches are juveniles of the severely endangered Kemp's Ridley Sea Turtle.

For 25 years Masachusetts Audubon and Boston's New England Aquarium have partnered to rehabilitate cold-stunned turtles and return them to warmer southern US waters.

It is crucial to recover these stranded turtles as quickly as possible. Do not assume that a turtle is dead—turtles that appear lifeless are often still alive.

If you come across a stranded sea turtle on the beach, please follow these simple steps:

1. Move the turtle above the high tide line. Never grab or hold the turtle by the head or flippers.

2. Cover the turtle with dry seaweed or wrack.

3. Mark the turtle with an obvious piece of debris—buoys, driftwood, or branches.

4. Call the Wellfleet Bay Wildlife Sanctuary hotline: 508-349-2615, ext. 6104.

Sea turtles are federally protected under the Endangered Species Act; as such, it is illegal to harass sea turtles or transport them without a permit, so do not try to move the turtle off the beach yourself.

Sharks. Though beautiful, Blue Sharks can be aggressive around bloody fish cleanings, earning them the nickname "blue dogs" among fishers.

Sea turtles

Although they are seldom seen except when injured or dead, the Outer Lands coastal waters are regularly visited by four sea turtle species, usually in late summer or early fall. The Green Turtle, Kemp's Ridley, Loggerhead Turtle, and huge Leatherback likely ride the Gulf Stream north into New England waters. The closest major nesting areas for Green and Loggerhead Turtles are found on Florida's beaches. The severely endangered Kemp's Ridley nests primarily on one small Gulf of Mexico beach in the Mexican state of Tamaulipas, just south of the US-Mexican border, and on other scattered locations in Tamaulipas.

Healthy sea turtles are creatures of the deep ocean and offshore coastal waters, and they normally come close to shore only to breed. No sea turtles breed in the region's relatively cool waters, and a healthy sea turtle only occasionally comes to the water's surface or shows much of its body out of the water when at the surface. As a result, even though sea turtles are present every year, healthy individuals are seldom noticed, even by experienced naturalists.

Unfortunately, many sea turtles that ride the warm Gulf Stream north into the region's much colder waters become cold-shocked and listless, particularly in late fall and early winter. In recent years there has been an increase in fall turtle stranding on beaches, perhaps because the local waters have warmed, allowing the turtles to be active longer into the fall. The chilled and helpless turtles often wash up on the beaches of the Outer Cape, where local naturalists from the Massachusetts Audubon Society and other organizations have organized efforts to collect the stunned turtles, allow them to revive in warm indoor shelters, and then fly batches of the turtles back to warmer waters in Florida. Although the effort is considerable, all sea turtles are endangered species, so every individual saved really counts.

Green Turtles and Kemp's Ridleys

Small numbers of mostly juvenile Green Turtles and Kemp's Ridleys appear in New York and southern New England ocean waters. These smaller sea turtles feed on bottom crustaceans like crabs and lobsters, and although the two species are primarily warm-water animals, the mid-Atlantic Coast appears to be an important feeding area for the young of both species.

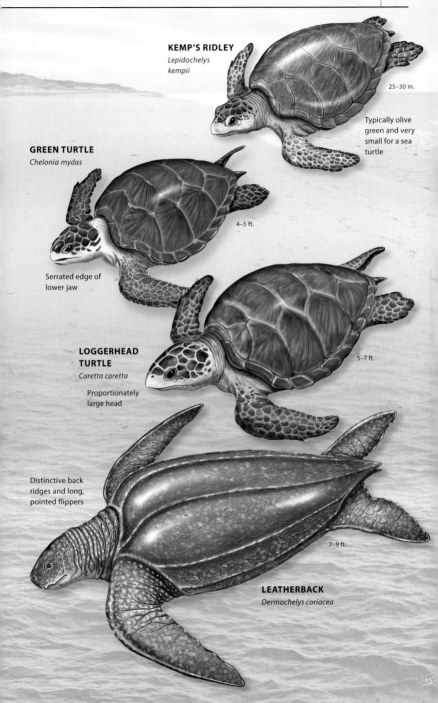

KEMP'S RIDLEY
Lepidochelys kempii

25–30 in.

Typically olive green and very small for a sea turtle

GREEN TURTLE
Chelonia mydas

4–5 ft.

Serrated edge of lower jaw

LOGGERHEAD TURTLE
Caretta caretta

Proportionately large head

5–7 ft.

Distinctive back ridges and long, pointed flippers

7–9 ft.

LEATHERBACK
Dermochelys coriacea

Loggerhead Turtle

The Loggerhead is the most common and robust of the smaller sea turtle species and appears in small numbers mostly south of Cape Cod. Loggerheads feed on a wide variety of bottom crustaceans, lobsters, and crabs. These sea turtles nest in small numbers as far north as the New Jersey coast, but most East Coast Loggerhead nesting takes place on Florida's Atlantic Coast and the Gulf of Mexico coast. Like all sea turtles, the Loggerhead is endangered, primarily by the loss of nesting habitat and by nest and egg disturbance. In Mexico, Loggerhead eggs are collected for food. Sea turtles also die in large numbers in commercial fishing operations and, even more tragically, when they are entangled in drifting abandoned fishing gear and drown because they can't reach the surface to breathe.

Leatherback

Probably because of their large size and distinctive ridged back Leatherbacks are the sea turtle species most often spotted in local waters. Leatherbacks can reach 8 feet in length, and a large individual can weigh 2,000 pounds. Off the ocean coasts of Cape Cod and Long Island deep-sea fishers and whale watchers often spot Leatherbacks at the surface. These powerful swimmers can range over oceans and dive more than 1,000 feet underwater in search of their main prey, sea jellies. Leatherbacks are unusual for turtles in that they are at least partially endothermic (warm-blooded) and can remain

Common Eider, male
Somateria mollissima

active even in very cold waters.

Leatherbacks and other sea turtles are often killed when they mistake discarded plastic bags or party balloons for sea jellies. The plastic clogs the digestive system, and the turtle starves to death or dies of abdominal infections. Never throw away plastic bags on a beach or from a boat and never deliberately release a party balloon, because these long-lasting items can be lethal to many kinds of marine animals, not just sea turtles.

All sea turtle species are on the New York, Rhode Island, Massachusetts, and federal endangered or threatened lists and should be neither approached closely by boat or on the beach nor handled in any way except by qualified experts.

Birds

The whole Outer Lands region is an important habitat for all kinds of shorebirds and water birds, and even some true seabirds. In summer the Outer Lands offer birds bountiful food and habitat that is relatively protected from the stronger winds and higher seas of the true ocean coasts to the east and south. In other seasons the region remains a rich and varied source of food not just for overwintering regional birds but also for East Coast migrants in spring and fall.

Most so-called seabirds that frequent coastal waters are really more shoreline birds—that is, they specialize in feeding along the immediate shore or the subtidal waters just offshore (see the previous chapter, "Shallow Coastal Waters"). The birds considered in this chapter regularly feed well offshore in the deeper waters, although some species such as cormorants and gulls feed in both the shallows and the depths. See the earlier chapter "Beaches" (pp. 144–61) for references to common

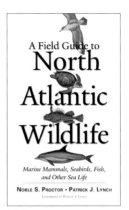

For more extensive coverage of offshore animals of the North Atlantic, see *A Field Guide to North Atlantic Wildlife*, also from Yale University Press.

Black Scoter
Melanitta americana

White-Winged Scoter
Melanitta deglandi

Surf Scoter
Melanitta perspicillata

GREAT CORMORANT
Phalacrocorax carbo

First-year
immature

Dark throat,
light belly

First-year
immature

Light throat,
dark belly

**DOUBLE-CRESTED
CORMORANT**
Phalacrocorax auritus

Adult

Imm.

Double-crested
in flight

Imm.

Uphill angle
of flight

Adult

Great Cormorant
in flight

**GREAT
CORMORANT**

**DOUBLE-CRESTED
CORMORANT**

Double-Crested
and Great adults in
breeding plumage

Outer wings are black to the wrist, unlike the more limited black wingtips of gulls

First-year juvenile

Second- or third-year immature

Adult plumage (at least five years old)

NORTHERN GANNET
Morus bassanus

The heavy bill, shielded nostrils, and thick head plumage are adaptations to diving from great heights into the sea

Adult

Herring Gull
Larus argentatus

Maciej Olszewski

Great Black-Backed Gull
Larus marinus

eugenesergeev

John Sandoy

Outer Lands shorebirds.

Cormorants

Over the past 30 years the Double-Crested Cormorant has expanded its year-round presence along New England shorelines, and this diving bird now breeds in many coastal places as well as on inland lakes. The larger Great Cormorant breeds along the coasts north of Cape Cod and is a regular sight in the Outer Lands in winter. Both cormorant species are strong fliers and expert fish catchers, and they will go well offshore to feed on schooling fish or squid.

Gulls

Most gulls are shoreline birds and do not typically wander far offshore. However, Herring Gulls, Great Black-Backed Gulls, and Laughing Gulls will fly considerable distances offshore to swarm fishing boats and feeding groups of Humpback Whales, where the gulls hope to pick off dead or injured fish.

Sea ducks

Outer Lands waters host large numbers of diving bay ducks such as scaup that primarily feed in the subtidal waters and typically flock within a half-mile of the shoreline, but a few species of wintering ducks routinely venture well offshore to deeper, rougher waters. These are the most marine species of ducks.

The Long-Tailed Duck (formerly called Oldsquaw) is known for the deepest, most sustained dives of any diving duck and has been known to descend as deep as 200 feet to forage for the clams and other marine invertebrates it favors. Sadly, these hardy ducks are now threatened worldwide, but southeastern New England still has a significant population of Long-Tailed Ducks.

Scoters are large, sturdy, black sea ducks, distinguished mostly by their head coloring. They are powerful underwater swimmers and feed primarily on mollusks. Three species of scoter overwinter in the Outer Lands coastal waters and are often seen well offshore over shallow shoals. The most common wintering scoter in these waters is the White-Winged Scoter, and some Surf Scoters are present throughout winter.

In winter off the Cape and Islands and off Montauk Point large rafts of Common Eiders are seen regularly, and these very large, Arctic sea ducks are best adapted to feed on shoals in ocean waters and are the most truly marine of all sea ducks. Common Eiders occur in huge numbers throughout offshore waters in the region, particularly off Nantucket and in Nantucket Sound.

Gannets

Northern Gannets are large seabirds that nest all around the Northern Atlantic, generally on the edge of rocky cliff sides. In the cooler months of the year gannets roam widely through the coastal and oceanic waters, including many areas off the Outer Lands.

Gannets make spectacular dives from 50 to 100 feet in the air, folding in their wings before plunging headlong deep below the surface and raising a large splash column that can be seen from a mile or more away. Feeding gannets are unmistakable; no other large, white seabird makes that kind of headfirst dive into schools of fish. Gannets are able to withstand the impact of hitting water from great heights through their streamlined heads and bodies, very thick feathers around the head and neck, and sturdy conical bill with specialized nostril openings resistant to the extreme water pressure of dives.

Gannets are present in small numbers offshore of the Outer Lands throughout the year but are especially common in the spring and fall.

Oceanic seabirds

Deep-ocean bird species such as dovekies, puffins, murres, razorbills, shearwaters, and jaegers all regularly occur off the ocean coasts of the Outer Lands, but these truly marine species seldom come within sight of land unless they are forced in by strong onshore winds or by ill-health. In strong nor'easters the counterclockwise circulation sometimes sweeps large numbers of seabirds into Cape Cod Bay, where

Laughing Gull
Larus atricilla

Dovekies fly in loose flocks

All shown in their winter plumage

Puffins and guillemots fly in lines

DOVEKIE
Alle alle

BLACK GUILLEMOT
Cepphus grylle

ATLANTIC PUFFIN
Fratercula arctica

GREAT SHEARWATER
Ardenna gravis

Crisp black cap; smudge on belly

CORY'S SHEARWATER
Calonectris borealis

WS 44 in.

Cap is smudgy and gray. Belly is uniformly light. Uncommon.

WS 46 in.

SOOTY SHEARWATER
Ardenna grisea

Uniformly dark, with gray wing linings. Smaller than Cory's or Great Shearwaters.

WS 40 in.

MANX SHEARWATER
Puffinus puffinus

Small, fast. Best field mark is the flickering contrast of the dark back and light underside.

Light wing linings

WS 32 in.

PARASITIC JAEGER
Stercorarius parasiticus

Uncommon. Jaegers are predatory gull relatives. They look like gulls but fly much more like falcons, swift and direct. Will harass other seabirds for food.

WS 46 in.

WILSON'S STORM-PETREL
Oceanites oceanicus

WS 15 in.

Not to scale. Storm-petrels are tiny seabirds that will often hover just above the surface when it's calm, dangling their feet into the water surface to attract krill and small planktonic animals.

they wreck on the bay shore of the Cape, forced onto the beaches or even inland by wind. When storms or blustery weather bring strong onshore winds, prominent ocean-facing locations such as Race Point, Nauset Beach, the eastern shores of Nantucket, Montauk Point, and the beaches of the south coast of Long Island are good places to look for seabirds from shore, particularly if you have a spotting scope.

Fortunately, the Outer Lands region hosts many seasonal whale watching operations that run large boats offshore from late in April through late October, and aside from the wonderful spectacle of whales, these boats are also the most practical places to view seabirds at such locations as Stellwagen Bank north of Cape Cod, northern Cape Cod Bay, and along the Great South Channel east and south of Nantucket. Dovekies, Atlantic Puffins, murres, and Razor-Billed Auks are regular and numerous visitors to the coastal waters of the Outer Lands, but unfortunately these small seabirds are most common in winter, when offshore boat trips are rare.

Shearwaters are true oceanic birds that are present during whale-watching season, and their numbers and large size make them easy to spot. Although shearwaters look superficially similar to gulls, they fly very differently, soaring fast and stiff-winged at wave-top height, looking for fish to pluck off the surface. Shearwaters are frequently seen near prime whale-watching spots, such as Stellwagen Bank, but are seldom visible from shore in normal weather. Both shearwaters and gull species that venture far offshore like the Herring Gull and Great Black-Backed Gull will flock to areas where Humpback Whales are feeding, hoping to pick off Sand Lance and other fish that are panicked or injured by the feeding whales.

GREAT SHEARWATERS
Ardenna gravis

Stellwagen Bank

Just north of the northern tip of Cape Cod, Stellwagen Bank is an underwater glacial outwash plain that rises above the deeper waters of the southern Gulf of Maine. The bank can be as shallow as 65–100 feet deep at its southern end, but the waters immediately east and north of the bank can be as deep as 800 feet. Since 1992 the bank has been the core of Stellwagen Bank National Marine Sanctuary, created to recognize and protect the unique and abundant marine life of the area.

The bank is famous for the concentration of marine life that gathers there, including many food fish species, as well as such marine mammals as whales, dolphins, and seals. Stellwagen Bank is a steep-sided plateau that lies in the path of strong, deep-water currents, including the Western Gulf of Maine Coastal Current as well as currents from the open Atlantic. As the deep currents sweep up the steep sides of Stellwagen Bank, they bring nutrients and minerals from the bottom up into the shallows of the bank, feeding the plankton and other marine life of the area, providing a rich and varied base to the food chain.

These waters are rich in two species that are critical to the larger fish and whales of the area: the Sand Lance and, for filter feeders like the Basking Shark and the North Atlantic Right Whale, the tiny copepod *Calanus finmarchicus* (see illustration, p. 397). Humpback Whales, Fin Whales, Common Minke Whales, dolphins, tuna, and Gray Seals all feast on the tiny but incredibly numerous Sand Lance of Stellwagen Bank. The many gulls and seabirds of the bank also feed on Sand Lance, either directly as predators or as scavengers of Sand Lance injured or panicked by feeding whales. Right Whales

A mixed flock of Great Shearwaters (*Ardenna gravis*) and Sooty Shearwaters (the smaller, darker birds, *A. grisea*) scatter before the bow of a whale-watching boat on the southern edge of Stellwagen Bank north of Cape Cod.

SOOTY SHEARWATERS
Ardenna grisea

Steve Byland
LONG-TAILED DUCK *Clangula hyemalis*

Natures Moments UK
LONG-TAILED DUCK *Clangula hyemalis*

Paul Reeves Photography
WHITE-WINGED SCOTER *Melanitta deglandi*

M. Carter
SURF SCOTER *Melanitta perspicillata*

Karen Popovich
GREATER SCAUP *Aythya marila*

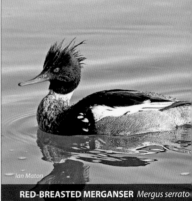

Ian Maton
RED-BREASTED MERGANSER *Mergus serrato*

and Basking Sharks feed on the huge swarms of tiny copepods that live in the waters of Cape Cod Bay and on Stellwagen Bank. Although individual copepods are tiny, the sheer volume of the animals in offshore waters is enough to provide the Right Whale and Basking Shark with the hundreds of pounds of food per day that they need to survive. Stellwagen Bank also supports Atlantic Cod, Yellow-Fin and Atlantic Bluefin Tuna, Silver Hake, Yellowtail Flounder, Striped Bass, and Bluefish, as well as sharks, including the Blue Shark and Great White Shark.

The great whales of Stellwagen Bank

Stellwagen Bank is visited nearly every year by a range of great whales and large dolphins, including the Blue Whale, Fin Whale, Sei Whale, Common Minke Whale, Orca (Killer Whale), Risso's Dolphin (Grampus), Long-Finned Pilot Whale, Atlantic White-Sided Dolphin, White-Beaked Dolphin, Common Dolphin, and Bottlenose Dolphin. Many of these species, however, enter the Gulf of Maine only sporadically in small numbers or in winter months when whale-watching trips are not available, so we'll concentrate on the three great whale species that are seen frequently on whale watches on or near Stellwagen Bank, Cape Cod Bay, and other areas off the Outer Lands.

North Atlantic Right Whale

The severely endangered North Atlantic Right Whale enters Stellwagen Bank and Cape Cod Bay waters in spring, as the Western North Atlantic population migrates north from the coastal waters off Florida, Georgia, and South Carolina, where they give birth to calves. The best time of year to see Right Whales in Outer Lands waters is late April, when whale-

The head of a North Atlantic Right Whale, showing its callosities (wartlike structures on the head and jaw). These skin growths are often swarming with whale lice that sometimes give the callosities a pink or orange tinge. Whale lice are not actually lice but small, shrimplike crustaceans that primarily eat algae and do not harm the whale. The average Right Whale hosts about 7,500 whale lice.

North Atlantic Right Whales
Eubalaena glacialis

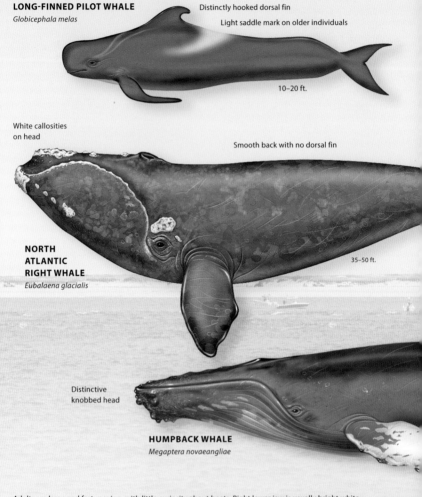

LONG-FINNED PILOT WHALE
Globicephala melas

Distinctly hooked dorsal fin

Light saddle mark on older individuals

10–20 ft.

White callosities
on head

Smooth back with no dorsal fin

**NORTH
ATLANTIC
RIGHT WHALE**
Eubalaena glacialis

35–50 ft.

Distinctive
knobbed head

HUMPBACK WHALE
Megaptera novaeangliae

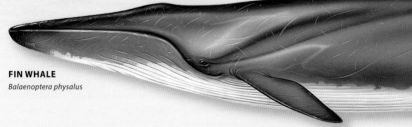

Adults are large and fast-moving, with little curiosity about boats. Right lower jaw is usually bright white. Back may show chevron patterns.

FIN WHALE
Balaenoptera physalus

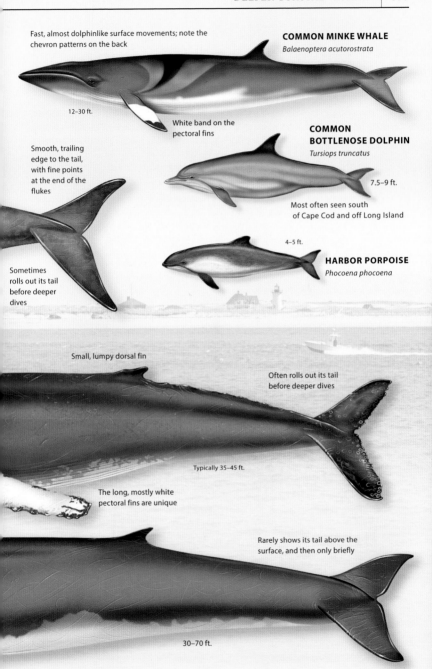

Fast, almost dolphinlike surface movements; note the chevron patterns on the back

COMMON MINKE WHALE
Balaenoptera acutorostrata

12–30 ft.

White band on the pectoral fins

COMMON BOTTLENOSE DOLPHIN
Tursiops truncatus

Smooth, trailing edge to the tail, with fine points at the end of the flukes

7.5–9 ft.

Most often seen south of Cape Cod and off Long Island

4–5 ft.

HARBOR PORPOISE
Phocoena phocoena

Sometimes rolls out its tail before deeper dives

Small, lumpy dorsal fin

Often rolls out its tail before deeper dives

Typically 35–45 ft.

The long, mostly white pectoral fins are unique

Rarely shows its tail above the surface, and then only briefly

30–70 ft.

watching fleets from Provincetown and other Cape Cod Bay ports begin their operations for the year. Because these whales are so endangered there are very strict limits for approaching them, so the views from whale-watching boats are often distant, but for dedicated whale watchers the trips to see one of the Atlantic Ocean's rarest animals is worthwhile. In 2016 about 440 North Atlantic Right Whales were counted in East Coast waters from central Florida north to the Gulf of Maine and Bay of Fundy. In the breeding season of 2018 researchers saw no live births of North Atlantic Right Whales, a shocking decline in an already endangered population.

As the spring weather warms the Right Whales drift north in the Gulf of Maine, to the shallow banks south of Nova Scotia, and into the Bay of Fundy. Cape Cod whale watchers have another chance to see Right Whales in late-season trips to Stellwagen Bank, in late October, as the Right Whales move south to their winter calving grounds off the southeast coast.

Humpback Whales

Each year the Humpback Whales of Stellwagen Bank make a 3,000-mile migration along the East Coast. Only in the early 1990s did scientists confirm that the whales we see off the Outer Lands migrate to winter and give birth to calves in the eastern Caribbean in the deep waters off the Greater Antilles (Cuba, Puerto Rico, and Hispaniola), the Lesser Antilles, and even as far south as Trinidad and Tobago, just off the

The massive gape of a feeding Humpback Whale (*Megaptera novaeangliae*). Humpbacks are able to dislocate their lower jaws at will (mandibular kinesis) to enable the maximum gape possible. The whale then closes its mouth, and uses its powerful throat muscles and tongue to force the hundreds of gallons of water out of the mouth through the baleen plates of the upper jaw, leaving hundreds of Sand Lance trapped inside the mouth.

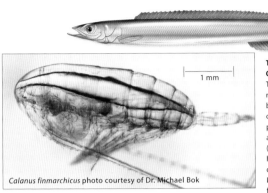

4–6 in.

1 mm

Calanus finmarchicus photo courtesy of Dr. Michael Bok

**THE FOUNDERS
OF THE FEAST**
Two small animals are
responsible for much of the
biologic richness and diversity
of the Outer Lands, and
particularly Stellwagen Bank
and Cape Cod Bay. Sand Lance
(top, *Ammodytes americanus*)
live in vast numbers in the
offshore waters of the Outer
Lands and are a critical food
for whales, seals, and many
other predatory fish and
birds. The tiny planktonic
copepod *Calanus finmarchicus*
is the primary food for two
large filter feeders, the North
Atlantic Right Whale and the
Basking Shark.

South American coast. As the weather warms in spring the
Western Atlantic Humpback Whale population moves up the
East Coast, probably using the Gulf Stream as a warm assist
as they move north. Many Humpbacks pass by the Outer
Lands area and continue on to the banks off the Canadian
Maritimes, Greenland, Iceland, and even northern Norway.
Luckily, hundreds of Humpbacks summer in the rich waters
of the Gulf of Maine and the Great South Channel just east of
Cape Cod and Nantucket. About 820 Humpback Whales were
reported in the Gulf of Maine and Bay of Fundy areas in 2016,
and the population of Humpbacks has been growing slowly
over the past decade.

In recent years small numbers of Humpbacks have summered in the New York Bight south of Long Island, and even
Long Island Sound and New York Harbor have had sporadic
reports of Humpbacks. It is thought that the new presence of
Humpbacks and other large whales in New York waters is an
indication of the growing population of Atlantic Menhaden,
an important forage fish species that has received attention
from conservationists and US government biologists.

An interesting fact: Orcas (Killer Whales) are seldom seen
in Outer Lands waters because the population is very small,
and scientists think that the Orca packs that do exist off the
East Coast favor deep ocean waters well offshore. However,
almost 15 percent of Humpbacks seen in the Gulf of Maine
show scars from Orca attacks, so the Killer Whales are out
there somewhere, perhaps north of the calving grounds in the
Caribbean.

A school of hundreds of
panicked Sand Lance struggle to
escape a bubble cloud moments
before a feeding Humpback
swallows the whole area in one
huge gulp.

Fin Whales
Fin whales are frequently seen by whale watchers in the Stellwagen Bank and Great South Channel and sometimes in the
waters south of Long Island, especially during spring and fall

migration. Like all the world's large whale species, Fin Whales are threatened, with about 1,600 individual Fins in the region from the mid-Florida coastline to the Bay of Fundy.

These days the primary threats to Fin Whales are collisions with ships, entanglement in fishing gear, and the overharvesting of forage food species like Atlantic Herring, Atlantic Menhaden, and Atlantic Mackerel. Fin Whales are known to undertake seasonal migrations, north in the warmer months and south as winter approaches, but the data on exactly where Fin Whales winter and calve are slim at present. Whale scientists are able to follow individual Fin Whales by their unique chevron markings, dorsal fin shape, and other physical details, which are unique in each animal. There is good evidence that the Fin Whales around Stellwagen Bank and the Outer Lands return year after year to the same feeding grounds.

Fin Whales are less well known among whale watchers because they show little curiosity or tolerance for boats, engage in few highly visible surface behaviors, and often leave the area quickly when whale-watching boats arrive. However, Fin Whales are not only the second largest animal on earth (only the Blue Whale is larger) but the second-largest animal *ever* to have lived on earth—far larger than any dinosaur or extinct marine animal. Fin Whales are huge and can travel at six to eight knots. Often the thrill of seeing these splendid but shy animals is witnessing the sheer size and grace of one of the ocean's most magnificent animals.

The 30-foot blow of a Fin Whale (*Balaenoptera physalus*) coming to the surface.

The rostrum (snout), splashguard, and blowholes of a Fin Whale individual known to researchers as Skeg. Also note the upper edge of the light-colored right jaw, a characteristic of Fin Whales. The left jaw of a Fin Whale is always dark-colored.

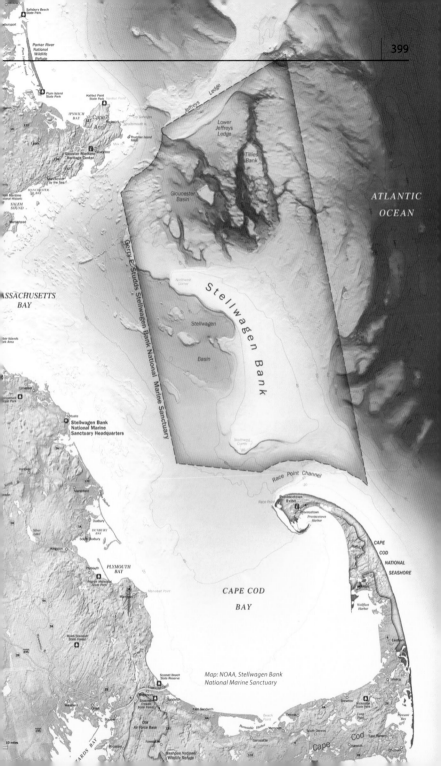

Map: NOAA, Stellwagen Bank
National Marine Sanctuary

NORTH ATLANTIC RIGHT WHALE (*Eubalaena glacialis*)
One of the rarest and most endangered animals on earth, with only about 440 individuals as of 2016. Right Whales typically occur in the offshore waters of the Outer Lands in spring and fall migration.

Typical size range: 35–50 feet

Average adult weight: 50 tons

Diet: Various zooplankton, in our area mostly the copepod *Calanus finmarchicus*

Approximate lifespan: 65–75 years

North Atlantic Right Whales will occasionally breach

Eduardo Rivero

Bonnet callosity

Blowhole

Eye

Pectoral fin

Habits: When feeding, this whale swims slowly at the surface with its mouth open. Often sluggish and surprisingly docile, sometimes suggesting the black, upturned hull of a sailboat more than a living whale. Can also be very energetic and acrobatic at the surface, breaching, rolling, tail-slapping, and spy-hopping. Often rolls its tail high above the surface when diving.

BLOW PROFILE
(as seen from behind)

V-shaped
and bushy

The foremost upper
callosity on the head is
called the bonnet

When coming to the surface to breathe, will
show much of the head and unique white
callosities

Right Whale head profile

Right Whale tail roll

When rolling at the surface, will
often wave pectoral fins in the
air before slapping them on
the water

Smooth back with no dorsal fin

Tail stock

Tail fluke

Right Whale

Humpback Whale

HUMPBACK WHALE (*Megaptera novaeangliae*)

Highly inquisitive, readily approaches boats. The Humpback Whale exhibits a wide range of sometimes spectacular social and feeding behaviors at the surface, including breaching, lob-tailing, and flipper-slapping. Humpbacks often feed in groups, where the whales cooperate to surround schools of small fish with nets or clouds of bubbles blown from their blowholes while underwater (bubble-netting). The blow is low and bushy. The Humpback often rolls its tail high out of the water at the beginning of a dive.

Typical adult size range: 35–45 feet

Average adult weight: 25–35 tons

Diet: Sand Lance, Atlantic Menhaden, Atlantic Herring, other smaller schooling fish

Approximate lifespan: 50 years

A feeding Humpback emerges at the center of a ring of bubbles it has blown (a bubble net) to trap a school of Sand Lance within the ring. The whale then surfaces in the center of the ring and gulps down the trapped fish.

Photo by Christin Khan, NOAA

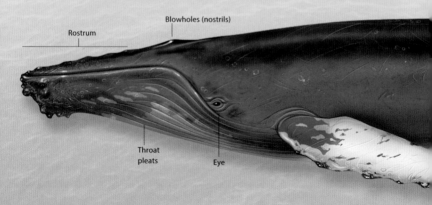

Rostrum

Blowholes (nostrils)

Throat pleats

Eye

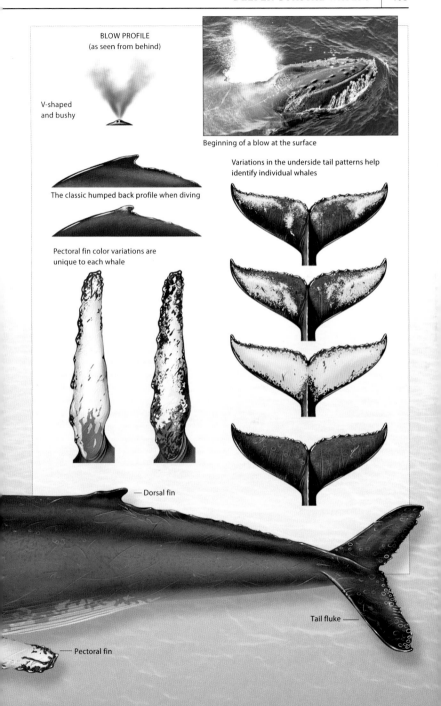

BLOW PROFILE
(as seen from behind)

V-shaped
and bushy

Beginning of a blow at the surface

The classic humped back profile when diving

Variations in the underside tail patterns help
identify individual whales

Pectoral fin color variations are
unique to each whale

Dorsal fin

Tail fluke

Pectoral fin

FIN WHALE (*Balaenoptera physalus*)
A sleek, fast swimmer despite its huge size. May approach drifting or very quiet boats but is indifferent to most vessels and apparently shy of engine noises. Sometimes performs low lunges across the water's surface while pursuing schools of fish; rarely breaches. Typical surface behavior is two to five blows, followed by a dive of five to ten minutes or more. Rarely rolls out its tail when diving. As its tail sweep up through the water, a submerged Fin Whale oftens leave huge circular footprints on the ocean surface, formed by upwelling water.

Typical adult size range: 30–7°0 feet

Average adult weight: 50–80 tons

Diet: Sand Lance, Atlantic Menhaden, Atlantic Herring, other smaller schooling fish, krill, squid

Approximate lifespan: 70+ years

Dark left jaw

Light right jaw; also note the chevron patterns on the back, behind the blowholes

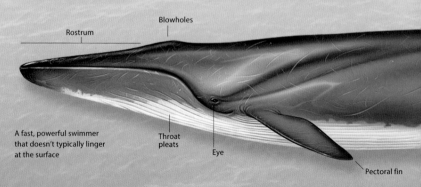

Rostrum

Blowholes

A fast, powerful swimmer that doesn't typically linger at the surface

Throat pleats

Eye

Pectoral fin

A typical whale watching view of a Fin Whale, powering away from the boat at great speed.

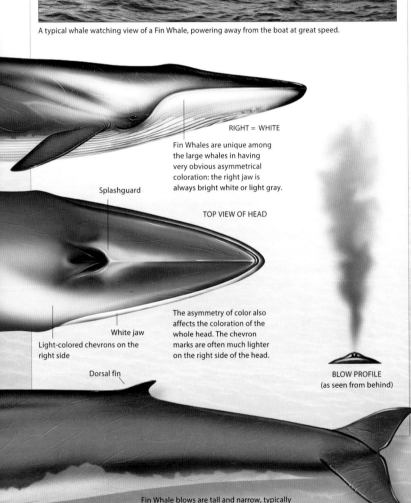

RIGHT = WHITE

Fin Whales are unique among the large whales in having very obvious asymmetrical coloration: the right jaw is always bright white or light gray.

TOP VIEW OF HEAD

Splashguard

The asymmetry of color also affects the coloration of the whole head. The chevron marks are often much lighter on the right side of the head.

White jaw

Light-colored chevrons on the right side

Dorsal fin

BLOW PROFILE
(as seen from behind)

Fin Whale blows are tall and narrow, typically 20–30 feet high. The noise of Fin blows can be heard long distances across open water.

Three Humpback Whales feed at the surface on Stellwagen Bank: a mother whale, her calf (in front), and an escort female companion of the mother whale. The birds are Herring Gulls, hoping to pick off dead or injured fish around the feeding whales.

SUMMARY OF SMALLER WHALES AND DOLPHINS

RISSO'S DOLPHIN (GRAMPUS)
Grampus griseus

Backs of older males become almost white with age and are heavily scarred

Relatively tall dorsal fin

Blunt, sloping head is never dark, as in pilot whales and most dolphins

8–13 ft.

Female shown; males are larger, with a more bulbous forehead

Female shown; male dorsal fins may be twice this height

ORCA (KILLER WHALE)
Orcinus orca

White eye and side patches are unique

Chevron patterns on back

Sharp snout often sticks above the surface when the whale blows

COMMON MINKE WHALE
Balaenoptera acutorostrata

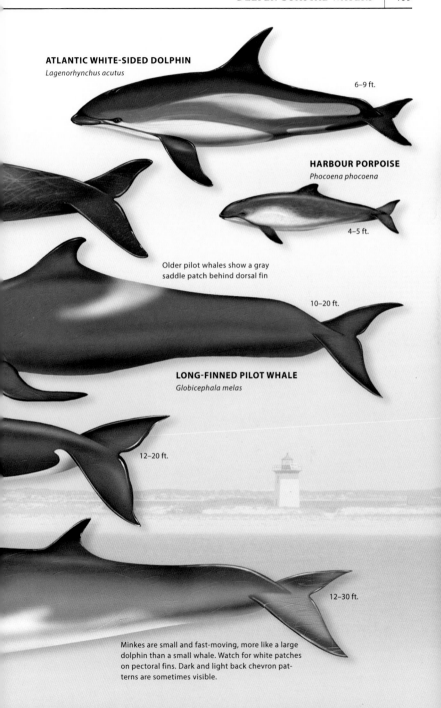

ATLANTIC WHITE-SIDED DOLPHIN
Lagenorhynchus acutus

6–9 ft.

HARBOUR PORPOISE
Phocoena phocoena

4–5 ft.

Older pilot whales show a gray
saddle patch behind dorsal fin

10–20 ft.

LONG-FINNED PILOT WHALE
Globicephala melas

12–20 ft.

12–30 ft.

Minkes are small and fast-moving, more like a large
dolphin than a small whale. Watch for white patches
on pectoral fins. Dark and light back chevron patterns are sometimes visible.

Race Point Beach, the Provincelands.

Further Reading

Books about the Outer Lands region

Allison, R. J. 2010. *A Short History of Cape Cod*. Beverly, MA: Commonwealth Editions.

Bellincampi, S. 2014. *Martha's Vineyard: A Field Guide to Island Nature*. Edgartown, MA: Vineyard Stories.

Beston, H. 1928. *The Outermost House*. New York: St. Martin's Griffin.

Cronon, W. 2003. *Changes in the Land: Indians, Colonists, and the Ecology of New England*. New York: Hill and Wang.

Cumbler, J. 2014. *Cape Cod: An Environmental History of a Fragile Ecosystem*. Boston: University of Massachusetts Press.

Dutra, J. J. 2011. *Nautical Twilight: The Story of a Cape Cod Fishing Family*. North Charleston, SC: CreateSpace.

Finch, R. 2017. *The Outer Beach: A Thousand-Mile Walk on Cape Cod's Atlantic Shore*. New York: Norton.

Philbrick, N. 2006. *Mayflower: A Story of Courage, Community, and War*. New York: Viking.

Philbrick, N. 2011. *Away Off Shore: Nantucket Island and Its People*. New York: Penguin Books.

Safina, C. 2011. *The View from Lazy Point: A Natural Year in an Unnatural World*. New York: Henry Holt.

Schneider, P. 2000. *The Enduring Shore: A History of Cape Cod, Martha's Vineyard, and Nantucket*. New York: Henry Holt.

Sterling, D. 1978. *The Outer Lands: A Natural History Guide to Cape Cod, Martha's Vineyard, Nantucket, and Long Island*. New York: Norton.

Thoreau, H. 2008. *Cape Cod: Illustrated Edition of the American Classic*. Boston: Houghton Mifflin.

An excellent site guide to Cape Cod

Schwarzman, B. 2002. *The Nature of Cape Cod*. Hanover, NH: University Press of New England.

General coastal and regional environments

Alden, P., and B. Cassie. 1998. *National Audubon Society Field Guide to New England*. New York: Knopf.

Bertness, M. D. 2007. *Atlantic Shorelines: Natural History and Ecology*. Princeton, NJ: Princeton University Press.

Finch, R. 1996. *The Smithsonian Guides to Natural America: Southern New England*. Washington, DC: Smithsonian Books–Random House.

Hay, J., and P. Farb. 1982. *The Atlantic Shore: Human and Natural History from Long Island to Labrador*. Orleans, MA: Parnassus.

Jorgensen, N. 1978. *A Sierra Club Naturalist's Guide to Southern New England*. San Francisco: Sierra Club Books.

Kaufman, K., and K. Kaufman. 2012. *Kaufman Field Guide to Nature of New England*. Boston: Houghton Mifflin.

Lippson, A., and R. Lippson. 1984. *Life in the Chesapeake Bay: An Illustrated Guide to Fishes, Invertebrates, and Plants of Bays and Inlets from Cape Cod to Cape Hatteras*. Baltimore: Johns Hopkins University Press.

Lippson, A., and R. Lippson. 2009. *Life along the Inner Coast: A Naturalist's Guide to the Sounds, Inlets,*

Rivers, and Intracoastal Waterway from Norfolk to Key West. Chapel Hill: University of North Carolina Press.

Lynch, P. J. 2017. *A Field Guide to Long Island Sound*. New Haven: Yale University Press.

Perry, B. 1985. *The Middle Atlantic Coast: Sierra Club Naturalist's Guide*. San Francisco: Sierra Club Books.

Proctor, N., and P. Lynch. 2005. *A Field Guide to North Atlantic Wildlife*. New Haven: Yale University Press.

Shumway, S. 2008. *Atlantic Seashore: Beach Ecology from the Gulf of Maine to Cape Hatteras*. Guilford, CT: Falcon Guides.

Weiss, H. M. 1995. *Marine Animals of Southern New England and New York: Identification Keys to Common Nearshore and Shallow Water Macrofauna*. Bulletin 115. Hartford: Connecticut State Press.

Beaches and dunes

Leatherman, S. P. 2003. *Dr. Beach's Survival Guide: What You Need to Know about Sharks, Rip Currents, and More before Going in the Water*. New Haven: Yale University Press.

Neal, W. J., O. H. Pilkey, and J. T. Kelley. 2007. *Atlantic Coast Beaches: A Guide to Ripples, Dunes, and Other Natural Features of the Seashore*. Missoula, MT: Mountain Press.

Shumway, S. 2008. *Atlantic Seashore: Beach Ecology from the Gulf of Maine to Cape Hatteras*. Guilford, CT: Falcon Guides.

Zim, H. S., and L. Ingle. 1989. *Seashore Life: A Guide to Animals and Plants along the Beach*. New York: St. Martin's.

Birds

Cape Cod Birding Club and Massachusetts Audubon Society. 2005. *Birding Cape Cod*. Yarmouth Port, MA: On Cape Publications.

Dunn, J., and J. Alderfer. 2011. *National Geographic Field Guide to the Birds of North America*. 6th ed. Washington, DC: National Geographic.

Olsen, K., and H. Larsson. 2003. *Gulls of North America, Europe, and Asia*. Princeton, NJ: Princeton University Press.

Peterson, R. T. 2010. *Peterson Field Guide to Birds of Eastern and Central North America*. 6th ed. Boston: Houghton Mifflin.

Sibley, D. 2014. *The Sibley Field Guide to Birds of Eastern North America*. 2nd ed. New York: Knopf.

Coastal forests

Jorgensen, N. 1978. *Southern New England: A Sierra Club Naturalist's Guide*. San Francisco: Sierra Club Books.

Sanford, G. R. 2013. *The Ecology of Woody Plants of Cape Cod*. Published by the author. ISBN: 978-1494485290.

Wessels, T. 1997. *Reading the Forested Landscape: A Natural History of New England*. Woodstock, VT: Countryman.

Fish

Boschung, H., et al. 1986. *The Audubon Society Field Guide to North American Fishes, Whales, and Dolphins*. New York: Knopf.

Budryk, P. 2014. *The Innermost Waters: Fishing Cape Cod's Ponds and Lakes*. Bloomington, IN: iUniverse.

Coad, B. 1992. *Guide to the Marine Sport Fishes of Atlantic Canada and New England*. Toronto: University of Toronto Press.

Robbins, C., and C. Ray. 1986. *A Field Guide to Atlantic Coast Fishes of North America*. Boston: Houghton Mifflin.

Geology

Leatherman, S. P. 1988. *Cape Cod Field Trips: From Yesterday's Glaciers to Today's Beaches*. College Park: Coastal Publication Series, Laboratory for Coastal Research, University of Maryland.

Oldale, R. N. 2001. *Cape Cod, Martha's Vineyard, and Nantucket: The Geologic Story*. Yarmouth Port, MA: On Cape Publications.

Sirkin, L. 1996. *Eastern Long Island Geology, with Field Trips*. Watch Hill, RI: Book and Tackle Shop.

Skehan, J. W. 2001. *Roadside Geology of Massachusetts*. Missoula, MT: Mountain Press.

Stahler, A. N. 1966. *A Geologist's View of Cape Cod*. Garden City, NY: Natural History Press.

Van Diver, B. B. 1985. *Roadside Geology of New York*. Missoula, MT: Mountain Press.

Insects

Borror, D., and R. White. 1970. *A Field Guide to the Insects of North America North of Mexico*. Boston: Houghton Mifflin.

Dunkle, S. 2000. *Dragonflies through Binoculars: A Field Guide to Dragonflies of North America*. New York: Oxford University Press.

Klots, A. 1951. *A Field Guide to the Butterflies of North America, East of the Great Plains*. Boston: Houghton Mifflin.

Mammals and other land animals

Conant, R. 1958. *A Field Guide to Reptiles and Amphibians of the United States East of the 100th Meridian*. Boston: Houghton Mifflin.

DeGraaf, R. M., and M. Yamasaki. 2001. *New England Wildlife: Habitat, Natural History, and Distribution*. Hanover, NH: University Press of New England.

Kays, R., and D. Wilson. 2002. *Mammals of North America*. Princeton, NJ: Princeton University Press.

Marine environments

Bertness, M. D. 2007. *Atlantic Shorelines: Natural History and Ecology*. Princeton, NJ: Princeton University Press.

Watling, L., J. Fegley, and J. Moring. 2003. *Life between the Tides: Marine Plants and Animals of the Northeast*. Gardiner, ME: Tilbury House.

Zim, H. S., and L. Ingle. 1989. *Seashore Life: A Guide to Animals and Plants along the Beach*. New York: St. Martin's.

Marine invertebrates and shells

Abbott, R. 1968. *Seashells of North America*. New York: Golden.

Gosner, K. 1978. *A Field Guide to the Atlantic Seashore from the Bay of Fundy to Cape Hatteras*. Boston: Houghton Mifflin.

Mienkoth, N. 1981. *The National Audubon Society Field Guide to North American Seashore Creatures*. New York: Knopf.

Marine mammals and turtles

Boschung, H., et al. 1986. *The Audubon Society Field Guide to North American Fishes, Whales, and Dolphins*. New York: Knopf.

Katona, S., V. Rough, and D. Richardson. 1993. *A Field Guide to Whales, Porpoises, and Seals from Cape Cod to Newfoundland*. 4th ed. Washington, DC: Smithsonian Institution Press.

Kinze, C. 2001. *Marine Mammals of the North Atlantic*. Princeton, NJ: Princeton University Press.

Leatherwood, S., and R. Reeves. 1983. *The Sierra Club Handbook of Whales and Dolphins*. San Francisco: Sierra Club Books.

Perrine, D. 2003. *Sea Turtles of the World*. Stillwater, MN: Voyageur.

Salt marshes

Roberts, M. F. 1971. *Tidal Marshes of Connecticut: A Primer of Wetland Plants*. Reprint Series 1. New London: Connecticut Arboretum.

Teal, J., and M. Teal. 1969. *Life and Death of the Salt Marsh*. New York: Ballantine Books.

Warren, R. S., J. Barrett, and M. Van Patten. 2009. *Salt Marsh Plants of Long Island Sound*. Bulletin 40. New London: Connecticut Arboretum.

Weis, J. S., and C. A. Butler. 2009. *Salt Marshes: A Natural and Unnatural History*. New Brunswick, NJ: Rutgers University Press.

Trees, plants, and wildflowers

Brown, L. 1979. *Grasses: An Identification Guide*. Boston: Houghton Mifflin.

Cobb, B., E. Farnsworth, and C. Lowe. 2005. *A Field Guide to the Ferns and Their Related Families*. Boston: Houghton Mifflin.

DiGregorio, M. J. 1989. *Cape Cod Wildflowers: A Vanishing Heritage*. Hanover, NH: University Press of New England.

Peterson, R. T., and M. McKenny. 1968. *A Field Guide to the Wildflowers of Northeastern and North-Central North America*. Boston: Houghton Mifflin.

Shuttleworth, F. S., and H. S. Zim. 1967. *Non-Flowering Plants*. New York: St. Martin's.

Sibley, D. A. 2009. *The Sibley Guide to Trees*. New York: Knopf.

Silberhorn, G. M. 1999. *Common Plants of the Mid-Atlantic Coast: A Field Guide*. Baltimore: Johns Hopkins University Press.

Tiner, R. W. 2009. *A Field Guide to Tidal Wetland Plants of the Northeastern United States and Neighboring Canada: Vegetation of Beaches, Tidal Flats, Rocky Shores, Marshes, Swamps, and Coastal Ponds*. Amherst: University of Massachusetts Press.

Wigly, N. and S. W. Carr. *Looking at Lichens: A Journey of Discovery Beginning on Cape Cod*. Brewster, MA: Cape Cod Museum of Natural History.

First Encounter Beach, Eastham.

Illustration Credits

All photography, artwork, diagrams, and maps are by the author, unless otherwise noted with a page credit and this listing.

Additional photography credits

xiv, Outer Lands Region from space, NASA; 4–5, Gulf of Maine bathymetry, Paul Illsley (www. paulillsley.com); 16–17, Ice field, NOAA; 18, Aquinnah cliffs, kozzi, CanStock; 37, Aerial of Gull Pond area, National Parks Service; 56, Gulf Stream, NASA Earth Observatory; 57, Sea surface temperatures, NASA Earth Observatory; 58, Algae blooms, NASA Ocean Color Group; 79, Champlain's map Nauset Harbor, Wikimedia Commons; 85, Fish flakes and fishing docks, Wikimedia Commons; 86, Georges Bank fishing, Library of Congress; 87, Georges Bank fishing, Library of Congress; 88, Provincetown fishing boats, Library of Congress; 89, Whaling ship *Wanderer,* Library of Congress; 95, Provincetown fishing trawler, Library of Congress; 96, Georges bank fishing, Library of Congress; 99, Steamship *Romance,* Library of Congress; 145, Rip current diagram, National Parks Service; 141, Sand Dollar, hereswendy; 186, Sunken forest, Fire Island, National Parks Service; 199, Grass Shrimp, Brian Gratwicke, Wikimedia; 263, Eastern Whip-Poor-Will, Patrick Comins; 335, Horseshoe Crabs, Frank Gallo; 363, Basking Shark, Greg Skomal; 368, Diatoms, NOAA Photo Library; 369, Marine invertebrates, NOAA Photo Library; 373, Moon Jellies, Hans Hillewaert; 375, Longfin Squid, NOAA Photo Library; 397, *Calanus finmarchicus,* Michael Bok; 399, Stellwagen Bank Map, NOAA; 402, Humpback bubble net, Christin Khan, NOAA.

Images used under license from Adobe Stock

13, Rocky Mountains, Videowokart; 47, Jones Beach area, viii; 62, Brant Point, Martin Lehotkay; 72, *Mayflower II,* Jim Curran; 97, Bourne Bridge, Christopher Seufert; 100, Woods Hole harbor, Paul Lemke; 136–37, Least Terns, Ray Hennessy; 140, Moon Jelly, Eddie Kidd; 140, Lion's Mane Jellyfish, Mady Rogers; 140, Sea Nettle, helgidinson; 221, Short-Eared Owl, Wenona Suydam; 257, Eastern Meadowlark, cratervalley; 258, Grasshopper Sparrow, Steve Byland; 262, Tiger Beetle, moneycue_canada; 262, Box Turtle, Brian E. Kushner; 262, Painted Lady, leekris; 262, Fowler's Toad, Brian E. Kushner; 263, Prairie Warbler, yoderphotography; 263, Pine Warbler, dmsphoto; 263, Carolina Wren, Brian E. Kushner; 263, Eastern Towhee, Charles Brutlag; 263, Spicebush Swallowtail, andromeda108; 265, Bobolink, David Watkins; 269, Icy pond, Maxal Tamor; 278, Green Darner, Riverwalker; 285, Spicebush Swallowtail, FotoRequest; 285, Tiger Swallowtail, underb; 286, Box Turtle, Brian E. Kushner; 286, Snapping Turtle, Arvind Balaraman; 286, Pickerel Frog, Steve Byland; 287, Muskrat, Tspider; 288–89, Freshwater pond, ead72; 289, Bullfrog, Tobias Arhelger; 316, Redstart, Paul Sparks; 318, Pine Warbler, dmsphoto; 320, RB Woodpecker, brm1949; 362, Humpback Whale, seb2583; 362, Great White Shark, Fiona; 363, Ocean Sunfish, Zacarias da Mata; 384, Herring Gull, Maciej Olszewski; 384, GBB Gull, eugenesergeev; 384, GBB Gull, John Sandoy; 385, Laughing Gull, enskanto.

Images used under license from Dollar Photo Club (now owned by Adobe Stock)

16–17, Glacier face, ALCE; 16–17, Taiga forest, Alex Yago; 16–17, Tundra, YuliaB; 21, Montauk Point, travelview; 133, Wineberry, Hilda Weges; 139, Piping Plover, Steve Byland; 140, Horseshoe Crab, Kevin Knuth; 190, Piping Plover, Steve Byland; 191, American Robin, Rachelle Vance; 191, Northern Cardinal, Dollar Photo Club; 191, American Goldfinch, Dollar Photo Club; 191, Red

Fox, Pim Leijen; 191, Meadow Vole, CreativeNature; 192, Raccoon, hkuchera; 192, Cottontail, man-dritoiu; 192, Skunk, Jimmy; 221, Wood Duck, wildphoto4; 222, Snapping Turtle, lightningboldt; 224, Greater Yellowlegs, Glenn Young; 225, Blue-Winged Teal, Steve Byland; 225, Gadwall, Steve Oehlenschlager; 225, Clapper Rail, pstclair; 230, Greenhead Fly, allocricetulus; 231, Harrier, Steve Byland; 237, Cottontail, mandritoiu; 281, Common Goldeneye, FeatherCollector; 281, Gadwall, Steve Oehlenschlager; 286, Green Frog, mayabuns; 287, Red-Winged Blackbird, Steve Byland; 287, Marsh Wren, Steve Byland; 310, Long-Eared Owl, lukicarbol; 316, Cooper's Hawk, Chris Hill; 316, Yellow-Rumped Warbler; 316, Yellow Warbler, Michael Hill; 318, Goldfinch, Steve Byland; 318, Tufted Titmouse, Paul Sparks; 318, Blue Jay, Canon Bob; 320, Robin, Rachelle Vance; 320, Downy Woodpecker, Greg Williams; 322, White-Tailed Deer, Nicolase Lowe; 322, Raccoon, hkuchera; 322, Deer Mice, DMM Photography Art; 322, Eastern Chipmunk, elharo; 322, Gray Squirrel, Orhan Çam; 322, Red Squirrel, Anterovium; 323, Flying Squirrel, Tony Campbell; 323, Red Fox, dannytax; 323, Coyote, Josef Pittner; 323, Cottontail, mandritoiu; 323, Woodchuck, Mario Beauregard; 323, Weasel, hakoar; 347, American Wigeon, Jim Shane; 347, Bufflehead, Erni; 347, Blue-Winged Teal, Steve Byland; 347, Brant, Marco Barone; 348, Goldeneye, feathercollector; 348, Greater Scaup, Karen Popovich; 348, Lesser Scaup, Steve Byland; 350, Gray Seals, stylefoto24; 351, Gray Seals, stylefoto24; 355, Harbor Seals, randimal; 356, Gray Seals, Mark Bridger; 362, Blue Shark, Fiona; 370, Sea Nettle, Gino Santa Maria; 371, Comb Jelly, Evan Travels; 373, PMOW Jelly, MSNN; 373, Sea Nettle, helgidinson; 390, Long-Tailed Duck, Steve Byland.

Images used under license from Shutterstock.com
129, Mole Crab, IrinaK; 188, Lone Star Tick, Melinda Fawver; 188, Black-Legged Tick, Sarah2; 189, American Copper Butterfly, anotherlook; 189, Mourning Cloak Butterfly, jps; 189, Opossum, Lisa Hagan; 221, Green-Winged Teal, Erni; 230, Lone Star Tick, Melinda Fawver; 230, Black-Legged Tick, Sarah2; 230, Dog Tick, Elliotte Rusty Harold; 257, Pipevine Swallowtail, Sari O'Neal; 287, Common Yellowthroat, ShutterStock; 320, Common Yellowthroat, Paul Reeves; 327, Osprey, Steve Bower; 347, Green-Winged Teal, Erni; 348, Surf Scoter, M. Carter; 348, White-Winged Scoter, Paul Reeves; 356, Harbor Seal, Wim Claes; 359, Blue Crab, Kim Nguyen; 359, Horseshoe Crab in Eelgrass, Ethan Daniels; 371, Longfin Squid, Jiang Zhongyan; 372, Cannonball Jelly, ymgerman 372, Cannonball Jelly, Sky2015; 372, Lion's Mane Jellyfish, Greg Amptman; 372, Lion's Mane Jellyfish, Konstantin Novikov; 372, PMOW Jelly, bottom, sciencepics; 373, Comb Jelly, John Wollwerth; 390, Long-Tailed Duck, Natures Moments; 390, White-Winged Scoter, Paul Reeves; 390, Surf Scoter, M. Carter; 390, Greater Scaup, Karen Popovich; 390, Red-Breasted Merganser, Ian Maton.

Highland Light, North Truro, Cape Cod's oldest and tallest lighthouse.

Provincetown's busy harbor.

Index

Dawn, Provincetown harbor.